Challenges in Indexing Electronic Text and Images

Challenges in Indexing Electronic Text and Images

Edited by
Raya Fidel, Trudi Bellardo Hahn,
Edie M. Rasmussen, and Philip J. Smith

ASIS Monograph Series

Published for the American Society for Information Science
by Learned Information, Inc.
Medford, NJ
1994

ISBN: 0-938734-76-8

Price: $39.50

Book Editor: James H. Shelton
Cover Design: Sandy L. Skalkowski
Book Design: Shirley Corsey

Printed in the United States of America

TABLE OF CONTENTS

Directory of Contributors

EDITORS

Raya Fidel (Section II)
Associate Professor
University of Washington, Graduate School of
Library and Information Science
133 Suzzallo Library, FM-30
Seattle, Washington 98195
Tel: 206/543-1888   Fax: 206/685-8049

Trudi Bellardo Hahn (Section I)
Director, Professional Development
Special Libraries Association
1700 Eighteenth Street, N.W.
Washington, DC 20009
Tel: 202/234-4700   Fax 202/265-9317

Edie M. Rasmussen (Section IV)
Associate Professor
University of Pittsburgh
School of Library and Information Science
620 LIS Building, 135 North Bellefield
Pittsburgh, Pennsylvania 15260
Tel: 412/624-9459   Fax: 412/648-7001

Philip J. Smith (Section III)
Associate Professor
Cognitive Systems Engineering Laboratories
Ohio State University
210 Baker, 1971 Neil Avenue
Columbus, Ohio 43210
Tel: 614/292-4120   Fax: 614/292-7852

AUTHORS

John A. Bailey (Chapter 9)
University of Tulsa
600 South College, Harwell Hall
Tulsa, Oklahoma 74104
Tel: 918/631-3148   Fax: 918/599-9361

Ronald L. Buchan (Chapter 10)
Lexicographer
NASA Center for AeroSpace Information
800 Landing Road
Linthicum Heights, Maryland 21090
Tel: 301/621-0103   Fax: 301/621-0134

Joseph A. Busch (Chapter 2)
Systems Project Manager
The Getty Art History Information Program
401 Wilshire Boulevard, Suite 1100
Santa Monica, California 90401-1455
Tel: 310/395-1025   Fax: 310/451-5570

Rebecca Denning (Chapter 12)
Research Assistant
Cognitive Systems Engineering Laboratories
Ohio State University
210 Baker, 1971 Neil Avenue
Columbus, Ohio 43210
Tel: 614/292-4120   Fax: 614/292-7852

Michael T. Genuardi (Chapter 11)
Senior Analyst
NASA Center for AeroSpace Information
800 Elkridge Landing Road
Linthicum Heights, Maryland 21090-2934
Tel: 301/859-5300   Fax: 301/621-0134

Donna Harman (Chapter 13)
National Institute of Standards and Technology
Building 225/A216
Gaithersburg, Maryland 20899
Tel: 301/975-3569   Fax: 301/975-2128

Susanne M. Humphrey (Chapter 8)
Information Scientist
National Library of Medicine
Bethseda, Maryland 20894
Tel: 301/496-9300   Fax: 301/496-0673

Wayne P. Johnson (Chapter 12)
Senior Associate Research Scientist
Chemical Abstracts Service
P.O. Box 3012, 2540 Olentangy River Road
Columbus, Ohio 43210
Tel: 614/447-3600   Fax: 614/447-3713

Lucinda H. Keister (Chapter 1)
Head, Prints and Photographs Collection
History of Medicine Division, National Library
of Medicine
Bethesda, Maryland 20894
Tel: 301/496-5961   Fax: 301/402-0872

Peter Liebscher (Chapter 6)
Associate Professor
Palmer School of Library and Information
Science, Long Island University
C.W. Post Campus
Brookville, New York 11548
Tel.: 516/299-2843   Fax: 516/626-2665

Lois Lunin (Chapter 3)
922 24th Street, NW
Washington, DC 20037-2229
Tel: 202/965-3924

Gary J. Marchionini (Chapter 4)
Associate Professor
University of Maryland
College of Library and Information Services
4121 Hornbake Library Building
College Park, Maryland 20742
Tel: 301/405-2053   Fax: 301/314-9145

Brij Masand (Chapter 15)
Scientist
Thinking Machines Corporation
245 First Street
Cambridge, Massachusetts 12142-1264
Tel: 617/234-1000   Fax: 617/234-4402

Nancy Mulvany (Chapter 5)
Bayside Indexing Service
265 Arlington Avenue
Kensington, California 94707-1401
Tel: 510/524-4195   Fax: 510/527-4681

Lorraine F. Normore (Chapter 12)
Senior Associate Research Scientist
Chemical Abstracts Service
P.O. Box 3012, 2540 Olentangy River Road
Columbus, Ohio 43210
Tel: 614/447-3600   Fax: 614/447-3713

June P. Silvester (Chapter 11)
Supervisor Abstracting/Indexing/Lexicography
NASA Center for Aerospace Information
800 Elkridge Landing Road
Linthicum Heights, Maryland 21090-2934
Tel: 301/621-0118 Fax: 301/621-0134

Philip J. Smith (Chapter 12)
(see above in *Editors*)

Stephen J. Smith (Chapter 15)
Scientist
Thinking Machines Corporation
245 First Street
Cambridge, Massachusetts 12142-1264
Tel: 617/234-1000   Fax: 617/234-4444

Dagobert Soergel (Chapter 7)
Professor
University of Maryland
College of Library and Information Services
4105 Hornbake Library Building
College Park, Maryland 20742-4345
Tel: 301/405-2037   Fax: 301/314-9145

David Waltz (Chapter 15)*
Vice President for Computing Research
NEC Research Institute, Inc.
4 Independence Way
Princeton, New Jersey 08540
Tel: 609/951-2700

Amy J. Warner (Chapter 14)
Assistant Professor
University of Michigan
School of Information and Library Science
550 East University, 304 West Engineering
Ann Arbor, Michigan 48109
Tel: 313/764-9376   Fax: 313/764-2475

* *Mr. Waltz was employed by Thinking Machines Corporation when the material for chapter 15 was first presented.*

PREFACE

Advances in a variety of technologies are rapidly raising new challenges and opportunities for the field of information science. In addition to online retrieval of bibliographic records and indexes for books, journals, and images, it is now possible to access databases containing the actual full-text, images and other media. Furthermore, because these text and multimedia documents are available online, new conceptual approaches for the exploration of these documents have evolved.

These changes, driven by the needs of users and by the possibilities opened up with new technologies, suggest a need to reconsider the nature of indexing and indexes. In addition, new technologies raise questions about the process of indexing.

The collection of papers in this volume represent an attempt to explore some of these issues. The first section, with papers by Keister, Busch, and Lunin, is concerned with indexing images. These papers consider principles for organizing images, and their implications for the design and use of access systems.

The second section, comprising contributions from Marchionini, Mulvany, Liebscher, and Soergel, focuses on new methods for navigating through documents that integrate text, images and other media. Under the general label of "hypermedia," consideration of this form of navigation raises questions about its relationships to traditional notions of indexing and about the process of indexing.

The third section, with papers by Humphrey, Bailey, Buchan, Silvester and Genuardi, and Smith, Normore, Denning and Johnson, further considers the process of indexing, more specifically the use of computer support tools to assist in this process. Tools used in current production environments are described, along with prototypes and concepts for future support systems.

The last section, consisting of discussions by Rasmussen, Harman, Warner, and Masand, Smith and Waltz on the indexing and retrieval of full-text documents, deals with indexing processes and their relationship to the information to be retrieved. The availability of full-text online provides access to a wealth of implicit knowledge to aid in search and retrieval. Developing techniques so that computers can effectively make use of this implicit knowledge is a major challenge confronting researchers today.

This collection of papers makes it clear that the field of information science in general, and the art and science of indexing in particular, remain very rich areas for research and development. Our hope is that these papers will point toward directions for answering some of the important questions confronted by the field, and serve to stimulate further discussion and research.

INDEXING AND ACCESSING IMAGES

INTRODUCTION

Trudi Bellardo Hahn

Paintings, sculpture, textiles, and other visual documents sans text have been a part of human history far longer than textual documents. Oddly, however, information systems to catalog and index visual documents have been developed much more recently than those designed to store and retrieve textual documents. The challenge of finding information in large collections of textual materials to support the work of researchers and answer questions has occupied library classificationists, descriptive catalogers, and indexers since at least the nineteenth century. Indeed, the complete history of indexing, abstracting, and cataloging systems for accessing written sources is several thousand years old—traceable to the earliest known Alexandrian libraries. Equivalent systems for accessing nontextual documents are relatively recent and, until a few years ago, comparatively primitive. Just within the last decade a small but intense group of librarians and information scientists has focused energy and thought on the problems of access to art works, photographs, prints, and other visual documents.

The three chapters in this section are based directly or indirectly on presentations delivered at the SIG/AH session entitled "Are There Universal Principles for Organizing Images?" held at the 1991 ASIS Annual Meeting. All three chapters address the issue of subject content, or "aboutness," in relation to the needs of the users of visual image information systems. Lucinda Keister's contribution is the closest of the three to the equivalent presentations at the SIG/AH session. Lois Lunin expanded her paper to include some content from her article "The Descriptive Challenges of Fiber Art" that appeared in *Library Trends*, 1990. Joseph Busch's role at the SIG/AH session was to react to the other presentations. For this volume, therefore, he incorporated some of those remarks, included some description of the Witt and Census projects that was in Catherine Gordon's paper at the same session, and then expounded his own theory of information retrieval systems appropriate to support the research of art historians. Busch organized his ideas and delivered the substance of the chapter in this volume at a conference held in Madras, India in August 1992.

Lucinda H. Keister heads the Prints and Photographs Collection at the National Library of Medicine. She and her staff answer thousands of queries each year about images on specific subjects for an interesting mix of users, including picture and publication professionals, health professionals, and others in the museum and aca-

3

demic communities. Before designing a new videodisc access system, Keister studied several years' worth of logs of user requests—questions that range from serious ("thalidomide babies") to silly ("people racing in wheelchairs"), and from simple ("photos of X-ray machines used in shoe stores in the 1950s") to complex ("images from the Renaissance which show that science and the humanities were closer together than they are today . . .").

Joseph A. Busch is the systems project manager at the Getty Art History Information Program. The users of the information systems he works with are primarily art historians, whose research questions never involve simple, unambiguous facts. As Busch explains in Chapter 2, the historian slowly and painstakingly collects and organizes multiple "facts" that depend on historical dimensions such as space, time, and point of view, and then interprets them to write history.

Lois F. Lunin works in two worlds: information science and art. Among her professional activities, she writes on imaging and multimedia for several publications, including *Information Today*, and lectures and consults on system design for organizations. In another world, she makes small fiber constructions, many of which have been shown in juried exhibitions regionally, nationally, and internationally. The perspective gained from creating her own art works, combined with the attitudes and knowledge from her experience as an information scientist, have led her to analyze the issues involved with designing an information system that incorporates images of fiber art. Lunin's analysis has information system users firmly in mind, especially as she describes the emerging vocabularies used by the professionals in the field—artists, exhibition juries, curators, art critics, educators, conservators, and historians. To meet their needs, a useful information system for fiber art objects must encompass the three-dimensionality of the documents—their composition, content, materials, design, and construction techniques, and, as well, the artists' intention.

All three chapters share a debt to Erwin Panofsky, the art historian and to Sara Shatford, who interpreted Panofsky's theories to the information community. Panofsky identified three levels of meaning in art works: pre-iconography (factual meaning), iconography (cultural meaning), and iconology (interpretive meaning). The first two of these he subdivided into *of* and *about*. These may be explained most easily by example. In a section of Michelangelo's painting on the Sistine Chapel ceiling, the image, in pre-iconographic terms, is *of* a reclining naked man. At the iconographic level, it is *of* a particular man, Adam. The first level requires little knowledge except the practical experience of the world. The second level requires the viewer to know something about a particular religion. But what is the picture *about*—that is, what is the underlying theme or intent? At the pre-iconographic level, the common human expression is perhaps "reaching," "longing," or "waiting." At the iconographic level, the subject is more obscure—again, knowledge of the religion is required to know that it represents the "creation of human life." The indexer can address these *of* and *about* aspects of the Sistine Chapel ceiling in creating subject access to an information system of art works. Even when the images are abstract or nonrepresentational, what they are *of* can be described and indexed in terms of colors, shapes, textures, size, materials, and techniques, as Lunin especially makes vividly clear in her discussion of fiber art.

Panofsky's third level, the iconological, interprets deep intrinsic meanings. It is best left to the art historians, theologians, or other scholars; iconology cannot be indexed with any degree of consistency.

This section sheds light on the issues of access to the visual documents addressed by each author—photographs, art works, and fiber art pieces. The principles discussed, however, will generalize to many types of visual documents that, with the rapid developments in computer technology, include the emerging computer-based graphic representations that are growing in variety and sophistication.

Chapter **1**

USER TYPES AND QUERIES: IMPACT ON IMAGE ACCESS SYSTEMS

Lucinda H. Keister

INTRODUCTION

User query data played an important role in the development of an automated still picture retrieval system at the National Library of Medicine (NLM). This chapter presents background information about the NLM collection and its users, describes typical user queries, and portrays representative queries. It identifies a particular picture query type, called the "image construct query," based on an analysis of user query data. The chapter also describes difficulties in handling image construct queries by existing conventional access systems, and it proposes improved cataloging strategy combined with picture surrogates as the most effective way to generate better image retrieval.

BACKGROUND

The Prints and Photographs Collection of the National Library of Medicine began in 1879 with the acquisition of a large group of engraved portraits of prominent persons in medical history. It evolved into a documentary picture collection housed in the Library's rare book section, and its holdings were made accessible in the 1950s through a conventional manual card catalog. The collection has the typical historical visual formats—from woodcuts to albumen photoprints. Some items are very precious, as in Durer's "Melencolia 1" (Figure 1); some ephemeral, but extremely valuable, as in the patent medicine card for "Cocaine Toothache Drops" (Figure 2); and some, quite frankly, seem completely devoid of artistic merit.

But even unartistic images are relevant for the collection's users, who come with a variety of needs and purposes and their own professional approaches and jargon. Records indicate that these users, who number about 2000 per year, are one-half picture and publication professionals and one-third health professionals, with the remainder divided among members of the museum and academic communities and the gen-

Figure 1. "Melancolia 1"

eral public. The images selected may appear in books, television documentaries, movies, or educational projects, or they may be used merely for reference.

Use of the collection is substantial, considering its specialized nature. The activity is manageable enough, however, in that the small staff has handled all routine functions including the busy reference, cataloging, and curatorial tasks. Consequently, staff members have a good grasp of how these activities affect each other.

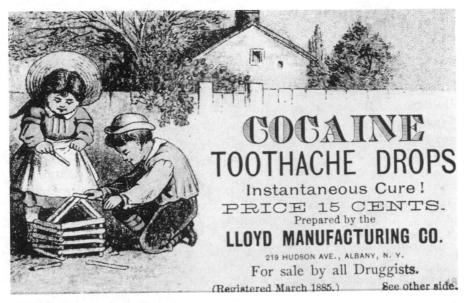

Figure 2. "Cocaine Toothache Drops"

USER QUERY DATA

The automated retrieval system for NLM's Prints and Photographs Collection is based on a detailed analysis of user queries. We documented those queries over the years in trying to find an answer to frustrating and unsuccessful reference interactions. Our method differs from a number of others' attempts to grapple with this problem that, although thoughtful and carefully considered, are essentially speculative.[1]

Subsequent analysis of our logs, in the context of possibly building a thesaurus, helped clarify and resolve problems that the staff encountered. These included the following: Why did so many pictures have to be retrieved from so many locales in the collection for one patron? Why could not patrons themselves find the pictures they wanted? Why could not staff find pictures they knew were in the collection?

Following the decision to implement a videodisc access system supported by MARC records in the Visual Materials Format, we again analyzed these logs but in a much more structured way. Reference orders and correspondence for one year (1984) were reviewed. Re-examining the reference query log allowed us to identify not merely terms, but the concepts that lay behind the seemingly casual, often amusing, user requests. It also helped us to understand why the existing card catalog had so little in common with the reference requests emanating from the "real world."

The review first showed that patrons do not ask for pictures in a consistent, traditional manner. Those patrons who are picture professionals (still picture researchers, TV, film, or media personnel) think visually and use art and/or graphics jargon; e.g., "an action shot of George Papanicolau, has to be horizontal and color." Health professionals ask for images in a manner more in keeping with NLM's orientation: "Do you have pictures of cholera?" The museum or academic community, on the other hand, often has precise citations to images it desires: " 'The Cow-Pock, or the Wonderful Effects of the New Inoculation' by James Gillray, please."

Unlike more controlled studies that canvass the query terms themselves, typically recording them verbatim, our quite pragmatic analysis forced us to reconstruct the queries from staffers' cryptic notes. As a result, we could not obtain hard, precise data on total numbers of queries and the subsequent proportional breakdown, such as numbers of requests for simple portraits compared to other requests. The goal was to discern conceptual differences revealed in the language of the queries themselves.

Staff members tried to reconstruct the query recorded in the log verbatim and, from memory, identify it with the user's profession. For example, an employee at Harry Abrams Publishing asked for "title pages of pharmacopoeias . . . botanical plants . . . 'Renaissance pictures' . . . digitalis . . . [and] 'serious pictures of the history of pharmacy.'" A science magazine writer requested pictures that were "old, like the 1930s, pictures that would give a sense of the marijuana controversy." A clothing designer specializing in authentic historical clothes wanted a picture of an early nineteenth century dentist and his family. A graphics firm needed "the HHS logo," and a frustrated TV producer wanted "polio patients between 1890-1914." Many more examples are shown in the Appendix. Our re-analysis was supported by the detailed logging of 239 additional queries during 1991.

QUERY PORTRAYALS AND DIALOGS

The three real-life reference interactions detailed below portray the process used to help patrons from varied backgrounds find the images they need.

Portrayal 1

An assistant editor for a public health journal was sent by her editor to find a "warm picture of a nurse, mother, and baby." Her supervisor had identified several images on the videodisc at the NLM exhibit booth the previous day that he thought she would select, including one that was "perfect." She duly reviewed each one, but none conveyed the *visual message* she had in mind. When staff retrieved all the pictures with "nurses" for her review (which at that time was the extent of our retrieval capability), she began to go through them one by one. She stopped at Number 89 because it struck her immediately as exactly what she wanted (Figure 3).

Portrayal 2

When the producer requested "polio patients between 1890 and 1914," staff replied that polio pictures in the collection were from the twentieth century. The patron responded, "Well, there were polio patients then . . ."

"Yes, I don't doubt it," the staff member said, "but I've never seen a photo identified as such in that time frame, which is not to say one doesn't exist, but you don't

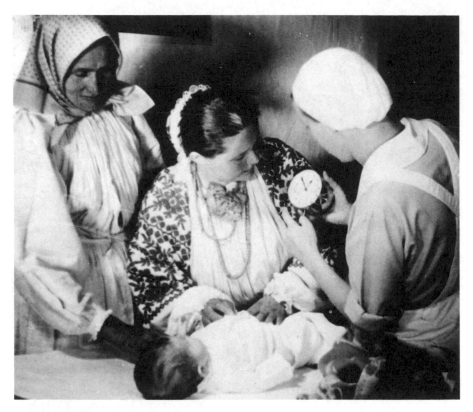

Figure 3. Photo of ". . . nurse, mother, and baby"

have the time to make that extensive search it will take to find it, do you?"

"No, I sure don't," he replied.

"Well, tell me how you visualize the image you want."

"Oh, poor people, especially children, maybe on a city street, lame or crippled, with canes." There it was, the query he had meant all along. It is doubtful that turn-of-the-century images ever have had the term "polio" associated with them, although it is likely many of the people in the pictures were indeed polio victims. If he could find it, he could use such a picture truthfully.

Portrayal 3

Dr. Michael DeBakey, the famed heart surgeon, once requested 20 specific images, processed as 8-by-10 glossy photos without copyright or museum restrictions—as soon as possible. One was a seventeenth century Dutch engraving of cats and dogs dressed as dentists and doctors (Figure 4). Did we have this or a substitute? He wanted what was obviously a fine print in the public domain, *if* we had it, but he had no hard word information to attach to it. The staff could not remember the picture. In fact, it

Figure 4. Engraving of cats and dogs dressed as dentists and doctors

comes up quite routinely in a search on the videodisc, but only because the thesaurus entry, "animals in human situations" from the Library of Congress's *Thesaurus for Graphic Materials* was attached to the record.

These examples provide a glimpse of the trench work required in still picture reference. Some observers may feel it unnecessarily intricate and that we do too much for our patrons. Nonetheless, the portrayals reflect the reality of picture research and show clearly the sometimes extremely difficult transition from verbal query constructions to retrieval of nonverbal information documents.

The difficulty is not only that the communication elements in an image are visual—another level of difference exists. Each of these documents acts as its own unique information vehicle, loaded with data, both obvious and subtle, all presented simultaneously in one neat visual rectangle. An audience member at a recent colloquium was struck by a speaker's description of the power of pictures: "He emphasized the differences between images and symbols, stressing that images are recognizable and are demonstrations of the things or concepts being presented. Image processing is faster than symbolic processing, and images are remembered longer and more easily. Images ground and connect people to the physical world, whereas symbols take them away from the physical world."[2] The contrast is similar to human linear or serial processing of words versus instantaneous pattern recognition of pictures. This important characteristic of still picture communication is a key to forming efficient approaches in cataloging and retrieval.

ANALYSIS OF QUERIES

The retrospective analysis of user queries identified the usual topical subject term requests: terms that also generally accompanied nonspecific visual requirements, such as "whatever you have on cholera in France in the nineteenth century is fine with me!" Some queries, however, can be more appropriately characterized as topic- or subject-independent. In these queries, patrons usually attempt to construct images with words, and these words define images that the user either remembers or invents to meet a particular immediate communication need (as in Portrayal No. 1 with the request for a "warm picture of nurse, mother, baby."). This type of query is well illustrated by how people request one of the most popular images in the collection, Benjamin Rush's "Tranquilizing Chair" for psychotic, maniacal patients (Figure 5). The image conveys with great simplicity and directness eighteenth century attitudes about insanity. It is generally asked for as the picture of "the man sitting in the chair with a box on his head."

Such image construct queries comprise one-third to one-half of the Prints and Photographs Collection requests. Although isolated terms in these queries may be topical, the concept behind them is a visual construct, e.g. "people racing in wheelchairs," or "surgeons standing" (requested by a surgeon to illustrate an occupational hazard for the profession). Often users describe the image contrary to its original intent and use it in an entirely different way. A very common example is the 1899 photograph of nurses in Figure 6, which was meant originally to convey forward-looking professionalism, but now is used to illustrate "nursing in the quaint old days." Such use is not manipulation of images—it simply illustrates that people use these visual rectangles at different times, in different ways, and for different purposes.

CATALOGING STRATEGY

Pictures cannot be retrieved in response to picture construct queries if the words that describe the picture (the "of" in "of and about"—the "factual" subdivision of Panofsky's "pre-iconography" level—or what one may also call the "visual elements") are inaccessible in the catalog record and/or in the search system. Sara Shatford's extensive discussion of still picture analysis for cataloging purposes sets forth a solid theoretical foundation for the task, but typically tends to overlook and underestimate the search potential of visual element access points within the catalog record.[3] Tipped off by our patron logs, we make sure the Prints and Photographs retrieval system incorporates visual element terms in item-level bibliographic descriptions of the pictures along with a surrogate of the still picture itself, i.e., the image on videodisc.

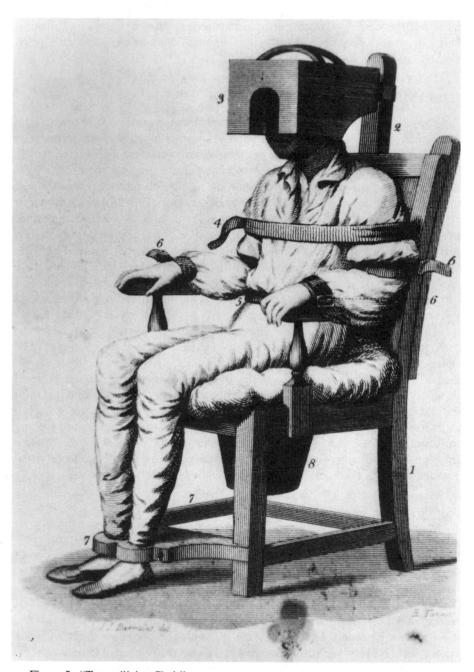

Figure 5. "Tranquilizing Chair"

Figure 6. Photo of nurses (1899)

The value of the standardized MARC Visual Materials Format catalog record is that it is nonidiosyncratic and transportable to other systems, predictable in its display and values, both in tagged and catalog card display. Prints and Photographs staff members produce brief, concise, and accurate item-level records meeting the needs of no more than 80 percent of the users, employing words in the record necessary only to describe the picture adequately. Added entries are chosen so as not to repeat words already in the body of the catalog record.

Because the host institution is a medical library, we add any information needed by the health professional, such as in the speculative diagnosis of Friedrich's ataxia in James Gillray's 1803 etching "The Burgess of Warwick" (Figure 7). In the main,

Figure 7. "The Burgess of Warwick"

however, the language selected is neutral, picking up on Elizabeth Parker's policy in *Graphic Materials*,[4] because neutral language meets the needs of most user groups. We make a sincere effort to use controlled language throughout the catalog record, taken from four thesauri: NLM's *MeSH*, the *Library of Congress Subject Headings*, LC's *Thesaurus for Graphic Materials*, and the Getty's *Art and Architecture Thesaurus*. For retrieval of all the image-construct queries, this effort has particular reward in the section of the MARC record we use to describe the visual content.

The catalog record for a still picture is thought of as having two parts—a sort of diptych. Each record contains both bibliographic information and a surrogate of the image itself, since in too many cases the selection process cannot be accomplished without seeing the actual image. Presentation of an image surrogate on a videodisc addresses the aesthetic or emotional need of the user—a highly subjective need not appropriate for the cataloger to consider. Any picture reference librarian knows that patrons ask for "dramatic pictures", "grabbers," etc. But what the cataloger/curator may think is a "mysterious" picture may not seem so to the graphics designer. Watching patrons searching images on the videodisc access system shows that a most interesting dynamic occurs in which words, in carefully constructed catalog records, introduce the user to selections of images; and the user then reviews, analyzes, and verifies with words again before finally arriving at the selection. The user constantly checks the television monitor to see which image "works," that is, communicates most effectively the desired message. Observing this dynamic in action validates the suspicion that still image research is a different critter entirely from standard search systems, which essentially retrieve words with words.

CONCLUSION

It is not so much that a picture is worth a thousand words, for many fewer words can describe a still picture for most retrieval purposes. The issue has more to do with the fact that those words vary from one person to another. As the examples in the Appendix indicate, words describing concrete image elements comprise a significant proportion of picture requests. Since these elements are more easily identified and can be agreed upon by various catalogers, paying close attention to this cataloging component can have very important benefits. It offers a higher level of precision in retrieval, even in collections that possess no image surrogate, since the terms at least indicate a picture's salient contents and help delimit the number of original items that users have to review.

A broader additional benefit is that visual element cataloging has the potential to increase access to still picture collections, nearly all of which are specialized, but usually contain many images of general interest, applicable to uses totally unrelated to the collection's main purpose. For example, a filmmaker seeking images of trained bears would not think to try the Prints and Photographs Collection at NLM—yet it has several such images as well as photos of sports, animals, plants, architecture, and many other categories that could be used by professionals far outside the health pro-

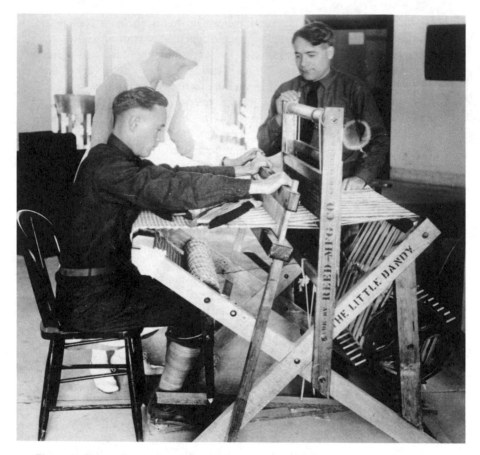

Figure 8. Photo of men weaving

fessional community. Lois Lunin [the author of Chapter 3] even found a photograph of men weaving (Figure 8).

While collection descriptions may adequately describe the scope of prints and bibliographic material, they often backfire in picture contexts. A better approach is to search directly among national automated picture databases that contain brief item-level catalog records with carefully selected access points.

ACKNOWLEDGMENTS

The author wishes to thank Craig Locatis, Ph.D., Lister Hill National Center for Biomedical Communication, for his kind, generous, and expert help in the editing of this manuscript.

NOTES

1. For two examples, see: Michael G. Krause, "Intellectual Problems of Indexing Picture Collections," *Audiovisual Librarian* 14, 2, May 1988; and "Hand-Waving computers," *The Economist*, July 27, 1991, 74.

2. Marilyn K. Moody, "More . . . On the Conference Circuit: The Imaging Revolution," *Technicalities* 10, 4, April 1990, 13.

3. Sara Shatford, "Analyzing the Subject of a Picture: A Theoretical Approach," *Cataloging & Classification Quarterly* 6, 3, Spring 1986, 39-62.

4. *Graphic Materials: Rules for Describing Original Items and Historical Collections*, compiled by Elizabeth W. Betz. Washington, Library of Congress, 1982, 5.

APPENDIX

The following is a selected list of user queries received by the Prints and Photographs staff from 1984 to 1991:

1. "that patent medicine card showing a girl peering around a column"
2. "people racing in wheelchairs"
3. "surgeons standing"
4. "faces in anguish"
5. "a warm picture of a nurse, mother and baby"
6. "a mysterious picture"
7. "that picture of Louis Pasteur with his granddaughter"
8. "a picture that shows chimps using sticks and poles to get bananas out of crates"
9. "imaginary scenes"
10. "pictures of black spots on white skin"
11. "the digestive system"
12. "19th century pictures showing female disorders, problems and weaknesses"
13. "a color image of an early laboratory engaged in working for the public good"
14. "fluoride"
15. "insomnia"
16. "photos of those X-ray machines used in shoe stores during the 1950s"
17. "solariums with wicker furniture"
18. "pictures of cradles"
19. "a soldierly scene"
20. "people using a plant to treat malaria"
21. "whooping cough"
22. "pictures of diseases that no longer exist"
23. "a 'cartoony' illustration showing how metastasis occurs"
24. "the Goddess Panacea"
25. "a picture of a 19th century doctor being very intimidating with a female patient"
26. "the 'diagnostic gaze' for a series of illustrations showing 'seeing' used in medicine: a) the diagnostic gaze, b) the microscope, c) the X-ray, and d) NMR and CatScans"
27. "a scary poster that a villainous dentist would use to show a patient what can happen to his teeth . . ."
28. "a picture that says 'rich resource of information'"
29. "that picture of the doctor sitting down beside a sick child" [Note: user meant Sir Luke Fildes' 1891 painting "The Doctor," widely reproduced and familiar Victorian painting]
30. "policemen"
31. "Nobel prizewinners"
32. "an older gentleman using an oxygen device"
33. "images from the Renaissance which show that science and the humanities were closer together than they are today . . ."

34. "pictures that show chiaroscuro"
35. "pictures of cows"
36. "lab coats"
37. "pictures in rich tones"
38. "that picture of the doctor, his horse, and saddle bags"
39. "syphilitic lesions"
40. "an action shot"
41. "they have to be vertical pictures"
42. "a scientist looking through a microscope between 1880-1900"
43. "the first X-ray—the one of Mrs. Roentgen's hand with a ring"
44. "neighborhood drug stores"
45. "intestines"
46. "hospital architecture through the ages"
47. "the cranium"
48. "historical use of animals in a lab"
49. "a patient taking a pill"
50. "consumption"
51. "pneumonia sign in a window"
52. "an edifice of medical learning"
53. "pictures that show alterations of appearance"
54. "a nineteenth century scene showing the population expansion of America"
55. "there were three big doorways in this picture . . ."
56. "old medicine bottles"
57. "a meeting of witches and demons"
58. "historical pictures of endocrine surgery"
59. "a soldier being bled"
60. "a person in pain from a toothache"
61. "cure of disease without drugs"
62. "birthing chairs"
63. "acupuncture charts"
64. "people preparing medicine"
65. "doctors reading books"
66. "a physician taking the history of a patient"
67. "nurses in emergency field service in World War II"
68. "midwives"
69. "Florence Nightingale surrounded by British politicians, cabinet ministers . . ."
70. "a woodcut of women attending to the dead body of a man"
71. "a picture of a front of a hospital, by itself—one that really looks like a hospital, especially to a child"

72. "New York hospitals in among other city buildings"
73. "medicine shows and quacks from the l930s"
74. "an 1890s photocomposition machine"
75. "photos of gravestones for missing limbs"
76. "an image that conveys the message 'bite the bullet' in a medical context"
77. "thalidomide babies"
78. "the Oath of Hippocrates"
79. "surgeons without gloves"
80. "a woodcut of tooth pulling"
81. "old Public Health Service poster"
82. "photographs of children's activities and children with single parents"
83. "Ishi-Hari charts for color blindness"
84. "acromegaly"
85. "pictures relating to the history of dining"
86. "mass burial grounds for American Indians"
87. "Irish pictures in your collection already on slides"
88. "aging and goat gonad therapy"
89. "chromosomes"
90. "frostbite"
91. "multiple births"
92. "Joseph Stalin's physicians"
93. "an entire hematology lab"
94. "ultraviolet treatment"
95. "using X-rays to treat acne"
96. "kidney stones"
97. "crude surgical instruments"
98. "image portraying psychotherapy when it was considered a stigma"
99. "attractive female doctor wearing glasses"
100. "small boy in a hospital bed"

Chapter 2

THINKING AMBIGUOUSLY:
ORGANIZING SOURCE MATERIALS
FOR HISTORICAL RESEARCH

Joseph A. Busch

INTRODUCTION

The Getty Art History Information Program (AHIP) has collaborated in developing an information system that contains information from the archives of three institutes for use in historical research. This chapter suggests ways to modify the conventional information retrieval (IR) model to accommodate an ambiguous IR response from historical sources. It reviews recent research on the cognitive process of inquiry and methods used by humanists, particularly historians and art historians, to organize information. A retrospective systems analysis approach describes and generalizes the Getty system data model. Examples of discoveries made using the Getty systems are briefly described.

DIFFERENCES BETWEEN HISTORICAL
AND FACTUAL INFORMATION RESOURCES

"More than one 'true interpretation' is indeed, very frequent in historical source material."[1]

Conventional information retrieval (IR) models are based on the premise that an answer to any question is *unambiguous*. Consider a simple example: what is the population of Los Angeles? An alphabetical search in the gazetteer in the *American Heritage Dictionary* under "Los Angeles" finds two places named Los Angeles—a city in Chile southeast of Concepción with a population of 49,175, and a city in southern California on the Pacific Ocean with a population of 2,966,763.[2] If the United States city were the *right* answer to the question, the dictionary (a factual database) would have supplied an unambiguous answer. A computerized gazetteer might have presented a list of multiple answers from which to choose, perhaps access to a map, and perhaps a variety of ways in which the query could have been posed (e.g., using string truncation and wild card characters). But the assumption informing the system design would state that a unique, unambiguous answer to the question exists, which could be formulated by typing in a string of alphabetic characters.

By contrast, an historical IR model is based on the premise that questions possess multiple and potentially *ambiguous* answers that depend on historical dimensions such as space and time, as well as points of view. Los Angeles (California) has existed over time, during which its political boundaries have changed. In an historical dictionary, Los Angeles would be variously identified as a part of Spain, Mexico, or the United States of America. It was not on, but near, the Pacific Ocean until the 1920s when the city incorporated several coastal regions. Population information would be represented by a series of counts derived from historical documents and censuses. A government demographer would report the population according to the latest U.S. Bureau of the Census data, while a community activist would include the uncounted homeless people and undocumented aliens. In an historical IR model, *all* of these population counts would correctly answer the question: "What is the population of Los Angeles?"

The scientist, technologist, businessman, public administrator, or student may wish to obtain a quick, unambiguous, and *correct* answer to a question. Many information resources such as almanacs, dictionaries, and their computerized counterparts have been designed to do just that. An information resource for the historian, however, requires a different sort of response reflecting a broader mode of inquiry. The resource should more closely resemble an entire library that contains a variety of types of information (primary, secondary, and tertiary) dispersed in many volumes, boxes, and shelves; and perhaps located in many different places throughout the world. The notion of historical research implies a mode of inquiry different from that used in the experimental sciences. It is difficult to predict what questions will be asked and what responses may be elicited. Not much can be said with certainty except that we live in the present and that there exist artifacts of the past. The historian's laboratories are libraries, archives, and museums, as well as the people, places, and objects found throughout the world.

The modern notion of history posits that no absolute or correct reading of the past exists, but only a subjective narrative informed by research into historical sources that are interpreted from the present point of view.[3] Simon Schama's recent book, *Dead Certainties*, provides two historical narratives, within which he reflects in the *Afterward* that "in both cases, alternative accounts compete for credibility."[4] What makes Schama's book modern is his self-consciousness about the variability of historical truth. Appropriately, parts of the book read like an historical novel, and others like a good mystery, serving to inform us of the narrow distinction between historical narration and fiction. Schama understands historical narrative as the telling of stories, not objective truths.

Today, Thucydides and Herodotus might more correctly be considered chroniclers, from the Greek *khronika* (annals, or a chronological record of events) rather than historians. As Schama implies, the modern notion of history concerns itself particularly with historiography (the writing and reading of history), the result of historical research, and its further interpretation and use over time.

REPORTS ON HOW HUMANISTS ORGANIZE INFORMATION

"History (historiography) is an inter-textual, linguistic construct."[5]

Helen Tibbo noted in a recent review article that a "realistic view of scholarship and information use" in the humanities is beginning to emerge[6] (p. 295). Recent studies of the process of inquiry and methods used by historians to organize and retrieve information show the importance and use of the following:

• Primary source materials in various formats and in large amounts,

• Proper names to organize and search for information generally, but

• Various heuristic methods to organize materials to be used in a specific project, such as chronologically or by topic, and

• Spatial metaphors to indicate the relationships between pieces of information.

Donald Case, reporting on a study of how historians organize their research materials in their offices, started with the notion that physical arrangements of both documents and office is of particular importance, since historians are obviously text-based researchers.[7] Case concluded that spatial metaphors in interface design for computer systems would be helpful. Cards and computerized lists were developed "to capture and sequence ideas and references to be used in writing a specific document" (p. 662), for example, to create detailed chronologies of what subjects had done during their lives. They also developed card files to the books and/or articles they had collected and read in support of their scholarship (p. 663). Most chose to organize their files and notes by topic, or by a combination of topic and chronology (p. 663). But "historians claimed to rely on their own memories for where a theme or particular passage can be found" (p. 664).

Marilyn Schmitt's interviews with art historians revealed the source materials that they find most important:[8]

• Visual materials such as original objects, reproductions, institutional collections of reproductions such as photographic archives, and personal collections of reproductions used for research and teaching;

• Primary documents such as original works in archives or dispersed, or reproductions and facsimiles of them; and

• Secondary sources such as books and articles in specialized subject libraries on university campuses, which are found using catalogs and indexes via printed or computerized access.

Schmitt noted the importance of documentation and cataloging information as a "creative stage in the process of intellectual inquiry" (p. 44). All interviewees kept files in a variety of media such as notebooks, card files, and data sheets. The "relationship of pieces of information plays a crucial role in conceptualization, even, at times, emulating spatial relationships" (p. 49). The most desired research enhancement is the creation of detailed scholarly catalogs with cross-indexing in various ways other than by artist, such as by iconography, time, region, and owner for information in museums, archives, libraries, slide libraries, and photo collections (p. 53).

Susan Siegfried and Deborah Wilde reported on an experiment to review computer searching behavior by humanities scholars at the Getty Center for the History of Art and the Humanities.[9] As in the other studies, a common theme was the importance of primary source documentary material. "The participants did a good deal of searching for proper names, either as subjects or as authors. This would seem to confirm Stephen Wiberley's contention that proper names form a larger part of the humanist's search vocabulary than has been recognized"[10] (p. 141). Wiberley also noted a strategy of some searchers who "panned for gold" from a printout of a large number of brief format records from which "they would select specific records to print [in] full" (p. 139).

Stephen Wiberley and William Jones, reporting on a general study of how humanists seek information, noted several distinct approaches, which include:[11]

• Geographic approaches in which "fellows who were studying the history or people of a locality usually went to that place to find evidence,"[12] and

• Genealogical approaches in which "much of the information seeking entailed tracking documents about individuals" (p. 643).

"We were impressed by how many fellows told us they did not talk to librarians who worked in general reference departments. In contrast, almost all fellows who used special collections, particularly of archives and manuscripts, reported that they depended heavily on the staff of these repositories" (p. 641).

Marcia Bates suggested that "area scanning," that is, by broad categories such as the way books are arranged in a library or documents in an archive, meets some real needs, although more research is needed to understand why this approach is popular.[13] The method often leads to serendipitous discoveries because the researcher can flip through pages of full-text materials to get "a quick gestalt" (p. 417). To provide area scanning in a database, it "will need to contain very large bodies of full text, as well as different types of text (narrative, statistical, bibliographic references, etc.)" (p. 419). Although Bates' study is not specific to the humanities, it does suggest a model of information-seeking behavior that appears congruent with those reported by studies that limited their scope to these disciplines.

Manfred Thaller reported on the historical workstation project at the Max-Planck-Institut für Geschichte in Göttingen.[14] He described the requirements of a system to hold and analyze both quantitative and textual information. The general requirements are for a structure that can accommodate data of variable length, a large number of attributes, and repeatable fields. The data may consist of full text in a small number of lengthy fields, structured data in many fields with very short texts, as well as data representing "temporal and numerical intervals" (p. 152). This project chose not to use a relational database model, but rather to develop a hypertext tool that permits the building of "book-style" databases. These databases can also be related to resources, such as currency tables, and have analytical and versioning tools available to support historical analysis and research.[15]

Some other recently developed computerized systems for historical source information include the Electronic Pierce Consortium,[16] which uses a hypertext tool to build a "book-style" database, the Medieval and Early Modern Data Bank (MEMDB),[17] which uses flat files, and the Kellogg Archive at Syracuse University Plexus system,[18] which uses a structured relational database system model.

GOALS OF THE GETTY SYSTEM DEVELOPMENT PROJECT

When the J. Paul Getty Trust, a not-for-profit foundation, was established in 1981, the staff decided to explore ways in which information technology could be applied to the problems of art historical research. Although many of the goals for the program were unclear, a particular interest was to create quickly resources that could be directly used by art historians. To accomplish this goal, several well-established research projects were funded to computerize their existing scholarly catalogs and cataloging operations. This chapter discusses two of these projects—the Witt Library Collection at the Courtauld Institute and the Census of Antique Art and Architecture Known to the Renaissance (the Census) at the Warburg Institute and Bibliotheca Hertziana.

A common information system was developed for the Witt Library and the Census. Its goal immediately diverged from that which initially motivated the Getty Trust. The system goal could be stated as follows: provide an improved method for accessing and navigating among the materials contained in the collections of the Witt Library and the Census, which have both visual and textual components.

The system needed to provide methods for Witt Library and Census staff to enter, edit, and control the consistency and quality of their information. In this sense, the system *users* were to be researchers who worked as information producers. The information had to be organized to conform to the pattern, style, and tradition of the Witt Library and the Census source information. At the same time these patterns had to be flexible enough to hold source information developed over the long histories of the respective projects and in their various subject areas.

The system also needed to provide a mode of access for researchers to search for and view information only, not to modify it. In this sense, the system *users* were to be researchers as information consumers. In this mode the researchers needed the ability to combine information in ways that the Witt Library and the Census could not have anticipated when they filled the databases, but not to change the information in the database. A method was also needed for the Witt and the Census periodically to generate new editions of the information system, as well as a method to replicate the system so that it could be distributed to additional sites.

BACKGROUND ON THE WITT AND CENSUS PROJECTS

Witt Computer Index (The Witt)[19]

The Witt Library Collection of the Courtauld Institute of Art in London contains more than 1.5 million photographic mounts covering Western painting from approximately 1200 to the present and assembled over the last eighty years. The collection is organized by School based on the nationality of the artists, then alphabetically by artist's name within School.

The Witt Library photographic mounts contain reproductions; printed text cut and pasted from published materials, such as museum, auction, and exhibition catalogs; and typed or handwritten annotations transcribed from published material or composed by the Witt Library staff. The text on the Witt mounts is called *catalogue raisonné* information. Catalogues raisonné identify a work of art by information such as artist, title, date, provenance (that is, ownership and sales history), exhibition history, and scholarship (that is, definitive sources of information about the artist or work). The computerized database records contain information directly transcribed from the mounts (with some notable exceptions described below). No effort was made to verify contradictory information that appears on the mounts.

The Witt Computer Index database currently contains information for about 57,000 works of art represented by approximately 67,000 photographic mounts for all artists in the American School in the Witt Library Collection and for about 12,000 eighteenth century works in the British School, and more than 160,000 associated authority records and 6,500 associated controlled vocabulary terms.[20]

Census of Antique Art and Architecture Known to the Renaissance

Since 1949, the Warburg Institute in London has collaborated with various scholars and institutes to develop a "central repository of scholarly findings about the relationships between the Renaissance and Antiquity."[21] The goals of the Census have been to identify and acquire (by purchase or photography) Renaissance sketchbooks after the Antique, drawings or copy studies made by other artists, and copies of existing photographs of ancient sculpture; implement photographic campaigns of ancient sculpture; and organize these source materials iconographically into documentary files "so that classical statuary and reliefs as well as their copies in artists' drawings and engravings can be seen in conjunction with medieval and Renaissance representations of the same subject." (*Annual Report, 1954-1955*).[22]

A 1986 inventory of the Census included a thematic catalog, a normal and an annotated index of artists, and an annotated illustrated index of Renaissance collections.[23] This inventory, an original scholarly work and not based on the computerization of the collection, reflected the unsystematic development of the collection over approximately 35 years.

The actual Census database is based on the information contained in the Warburg Institute's card catalog, which lists "all figured monuments, statues, sarcophagi, and bronze sculpture of Classical Antiquity that were known between 1400 and 1527, along with pertinent sources of texts and illustrations."[24] The database also includes Renaissance drawings and plans after the Antique, printed guidebooks of Rome, and facsimile manuscripts collected in the Bibliotheca Hertziana in Rome. The Census database currently contains approximately 43,000 computer records as well as 25,000 images stored on videodisc.

CHARACTERISTICS OF THE GETTY SYSTEM DATA MODEL

The initial Getty information system was developed by a third party, Online Computer Systems (Germantown, Maryland), using the standard systems development methodology—analyzing the user's requirements, translating the user's requirements into system terms, system design, system development, system testing, etc. When the Getty information system was designed in the early 1980s, however, "the whole field of systems analysis as a formal methodology, and specifically the database design using entity-relationship modeling and related techniques,"[25] were not widely in use.

Systems Analysis

In a recent interview, Dr. Catherine Gordon, the head of the Witt project, remembered the day she first met Rick Holt from Online Computer Systems at the Witt Index offices when they were located in Portman Square in London. She recalled that he had a large black briefcase with him. When they sat down at a table she asked him, "Well, what do you have?" He answered, "Well, what do you want?"[26] This interchange characterized the opportunity that this system development project presented for building an information system "from scratch."

Holt proceeded to base the information system model on an investigation and understanding of the cognitive patterns embodied in the information and its organization in the Witt Library Collection and Census documentary files. In the case of the Census, the system design was also based on an analysis of the card catalog at the Warburg Institute, supplemented by extended interviews with staff members working in each project.[27] This approach assumed that, given these two patterns of organizing information, a generalization from them might be valid more widely across the domain, and would also make it easier for outsiders to make more general use of the systems for research.[28]

Data Characteristics

A primary characteristic of the Witt and the Census catalogs was that all of the information was textual except for dates and images. The other key data characteristics were as follows:

- Extensive use of formal naming;
- Multiple sources of evidence, some of which *contradict* each other; and
- Indication of the relationships between materials.

Naming

The Witt and Census source materials make extensive use of proper names. These include the names of:
- Objects—such as works of art, monuments, and documents;
- People—such as artists, engravers, authors, wives, husbands, lovers, children, and students;
- Institutions—such as owners, exhibition organizers, professional associations, and dealers;
- Places—such as countries, states, cities, and buildings;
- Events—such as sales, exhibitions, and preservation actions.

These names may be repeated many times throughout the source materials, although frequently in many variant forms because of long development histories.

The Witt and Census source materials also use dating extensively, although dates were often named rather than stated explicitly (e.g., "mid-seventeenth century" or "Flavian") implying a range of dates rather than a specific one. In computerizing the Witt and the Census, a method was needed to search by dates as names, but also to define such names by starting and ending dates so that, for example, a date named "mid-nineteenth century" might be defined as between 1836 and 1862. On the other hand, when searching for an item between two dates, it might be important for the scholar to also be able to find items with date ranges that were partially within the specified dates. For example, in searching for "mid-nineteenth century," it might be important to also retrieve items with the dates named "nineteenth century" and "romantic period," etc. This concept of named date ranges is illustrated in Table 1.

The Witt and Census also used a variety of descriptive vocabularies to describe,

Table 1. Concept of named date ranges.

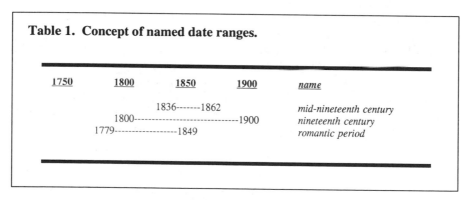

1750	1800	1850	1900	*name*
		1836-------1862		*mid-nineteenth century*
	1800---------------------------1900			*nineteenth century*
1779-----------------1849				*romantic period*

for example:
- Type of material—such as "oil on canvas" or "marble;"
- Visual medium—such as "drawing: pen" or "wash;"
- Type of institution—such as "town hall;"
- Role of an individual—such as "father;"
- Type of preservation action—such as "damaged" or "restored."

Some of these were small and not extensible; others were routinely extended as new descriptive terminology was needed. Appendix 1 lists all of the descriptive vocabulary and authority controlled fields with examples that are in the Witt and Census databases.

A special case of descriptive vocabularies involves descriptive details, or iconographical analysis. In art history, the examination and interpretation of visual information is extremely important. The methodology of iconographical analysis specifies three levels at which pictures can be interpreted:[29]

• The primary level describes the *natural* subject matter, which answers the question "What is it?" This may include the simple naming of more or less important objects in the primary sources, such as "female" and "cauldron."

• The secondary level describes the motifs identifying themes and concepts, which answers the question "What is it about?" For example, "Saint Cecilia martyred in a cauldron of boiling oil."

• The third level of analysis concerns interpretation, or the implied meaning of the object.

For these various types of *names* (i.e., names, dates, and descriptive vocabulary terms) in the Witt and Census, consistent access to the same *name* within the information system obviously was vital. Thus it was important to avoid typing in the same name, date, or term each time it was needed when cataloging or searching the database.

Historical Evidence

The Witt and Census source materials are concerned with collecting evidence or documentary materials relating to the works of art. Each piece of evidence is an historical event in the *life* of a work of art. The evidence considered together represents its history. Deirdre Stam proposed a "system for classifying evidence according to its historical and intellectual function."[30] The typology in Table 2 is based on Stam's model of art as communication. This typology is interesting particularly because the kinds of information compare closely to those in the Getty system model. It also illustrates that several pieces of historical evidence may document a particular historical event, and that many historical events may be associated with a particular work of art, as illustrated in Figure 1. One further aspect of historical evidence is that much of it is fragmentary or contradictory, and virtually all of it is subject to change[31] or reinterpretation. In many cases, the historical evidence may be ambiguous and thus open to many potential interpretations. In her most recent paper, Gordon gave an illustrative example in which the identification of the artist, title, subject, size, as well as the sitters vary among four pieces of historical evidence collected on Witt mounts for the same painting.

The same concept of historical events and historical evidence to describe them can be applied to names. Like objects, people, institutions, locations, and events also may have many historical events associated with them, as well as several pieces of historical evidence documenting those particular historical events.

Table 2. Typology for classifying evidence according to its historical and intellectual function. (Based on Stam)

Historical Events	Historical Evidence
artist's life	archives, correspondence, diaries, legal documents, biography
creation of the work	object, artist's drawings, diaries, correspondence, contracts
physical changes in the work over time	drawings, photographs, conservation records, scientific studies
society's reception of the work	guide books, sales records, inventories

Relationships

Whole-part relationships were implied by the way in which original source materials were being processed by the Witt and Census staff as well as by the scholars who were using them. Examples include the Witt's practice of tearing apart exhibition and sales catalogs to paste them onto mounts, or a scholar's reconstruction of a sketchbook whose drawings had been dispersed. Informed by an interest in the emerging field of artificial intelligence, Holt must have recognized that there were certain canonical patterns implicitly used by the projects for expressing the relationships between information about the works of art.[32] These were:

• Object to Evidence—one-to-many relationships between an object and its historical evidence as discussed above.

• Derivative Object—relationships from one object to one or more other objects to indicate that one was derived from another: for example, to reference copies or engravings to the source work of which it is a copy, replica, or engraving.

• Related Object—relationships from one object to one or more other objects to indicate a general reference to, for example, related works in the collection.

• Whole-Part—relationships from one object to one or more other objects to indicate that an object was part of another object(s).

• Part-Whole—relationships from one object to one or more other objects to indicate the parts of an object.

The first three of these types of relationships are nonrecursive, that is, only the explicit relationships are meaningful. For example, if A is related to B and B is related to C, it is not necessarily true that A is related to C. However, the whole-part and part-whole relationships are recursive into full hierarchical structures. It is also

32

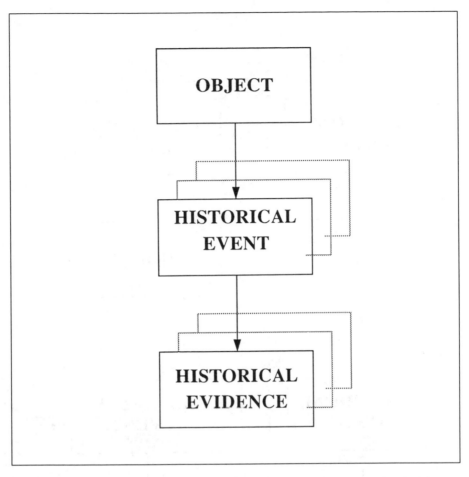

Figure 1. Diagram illustrating that one or more events may be associated with an object which is documented by one or more items of historical evidence.

possible that an object may be associated hierarchically to several different parents in a polyhierarchical structure; for example, a statue that was once a part of a Roman temple may have later been moved to a church, or the panels of an altarpiece may have been dispersed and later become associated with a different set of panels. Such mutable relationships are another consequence of the fragmentary and ambiguous evidence that exists, particularly for older objects such as Antiquities.

As with historical evidence, people, institutions, locations, and events may also have relationships between them; for example, the affiliations of artists with institutions such as professional associations, galleries, and museums and the dates of those affiliations, as well as their relationships with other individuals such as teachers and family. Objects, historical events, and evidence may also be associated with names (Figure 2).

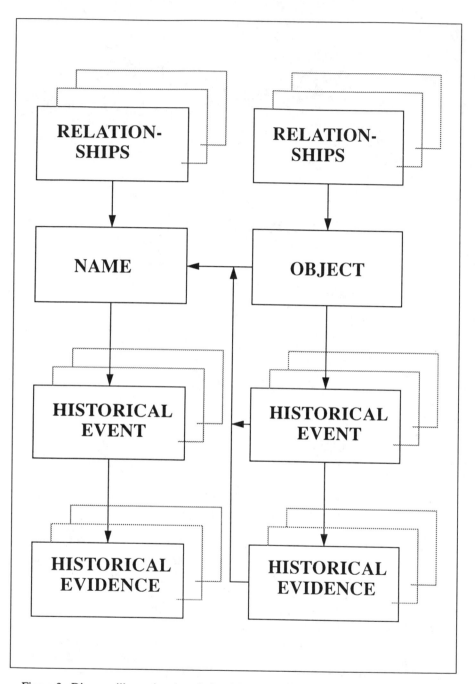

Figure 2. Diagram illustrating the relationships in the Witt and Census source materials.

Decomposition of Source Materials

The goal of the Witt and Census information system was not simply to access but also to navigate among the source materials. The problem was how to record the implicit relationships hidden within the source materials so that they could be accessed and used to navigate among them. The materials were also organized either physically together in files and, in the case of the Census, also cross-indexed intrinsically through the existing card catalog and extrinsically through the publication of scholars' work.[33] But it was decided that this system could not simply be an index to the source materials in the Witt and Census but would include the physical content of the sources as well.

An analysis of the information in the source materials in the collections showed that, even though they came from a variety of sources, they were relatively well-structured in a manner more or less consistent with the catalogue raisonné and other bibliographic models. Understanding the characteristics of the source information discussed above made it possible to decompose each type of source information into a group of uniquely occurring or repeating fields of information or data. Data entry would be a matter of deconstructing the text as it appeared in the source materials, and typing it into the appropriate data field.[34]

Source information for the various types of names that would occur in data fields also were decomposed into groups of fields according to the type of name. Once a name record was constructed, the unique name and its associated information would be stored only once in the database and simply could be referenced each time they were needed. Descriptive vocabularies were compiled more simply as finite or extensible vocabulary lists according to the type of vocabulary, which could be referenced each time a term was needed.

Finally, on the basis of the understanding of the relationships between the types of source information described above, the structure to reference or link between groups of fields was specified. Appendix 2 is an outline of the data structures of the Witt and Census databases.

Implementation of Systems

The information system embodying this information model was implemented in 1983. Two versions of the system were created—one for the Witt and one for the Census. They are nonetheless essentially the same. While minor enhancements to the initial system design have been made since then, the Getty system has maintained the original character of its data structure.

During the past two years, systems engineers from Digitus, Ltd. ported the system from its original minicomputer-based hardware platform to run on 386 microcomputers. A read-only version with simplified query-by-form screens was completed by Digitus in 1991. Initial testing and evaluation of the read-only system is in process. The new system has been replicated at the Witt, the Census, and Getty offices.

The Witt and Census research staff have been filling their databases with source information since 1983. While much information has been transcribed directly from source materials without interpretation, considerable research has been done in the process of constructing extensive authority files for names, institutions, locations, and dates. As noted above, the authority files also make it possible to specify certain relationships between them. Such information can provide insights into the provenance of objects through the history of their ownership, as well as the relationships between creators and their social milieu.

Although a considerable level of effort was required to build these authorities, over time the database has reached a critical mass so that the time saved by not needing to research and re-type names, institutions, locations, and dates is beginning to show dramatic productivity improvements. Furthermore, as the data structure is filled with the Witt and the Census source information, researchers using the databases have discovered unpredictable relationships between various data items.

DISCOVERIES USING THE WITT AND CENSUS SYSTEMS

The major advance of the Getty system is that, once one form of question has been posed, information responding to a range of other questions is immediately available without the need to formally pose these additional questions. For example, asking the question, "Where was this painting exhibited?" the user can find all other objects in the database that were included in the exhibition. In this way, it is possible to reconstruct an exhibition for which the catalog no longer exists (or ever existed). While the Witt Library has assiduously cut up the catalogs and reorganized their contents by artist, the database allows users to reconstruct them. As mentioned above, an interesting aspect of this information system is that relationships can be specified between both authority and nonauthority files; in browsing the database, users can move to any linked data.

Collecting dispersed information sources in one electronic resource has already led to some interesting scholarly investigations and discoveries. Some preliminary examples are as follows:

• The identification of an unknown sitter in a famous Joshua Reynolds oil sketch through the discovery that a distinctive piece of jewelry appears in this sketch as well as one in which the sitter is known. The identification of the diamond brooch was entered as part of the iconographic description of both works. The connection was discovered by searching on the ICONCLASS code describing the piece of jewelry.[35]

• The study of Raphael's archaeological method (he is, of course, much better known as a painter) suggests that the artist applied various architectural drawing techniques much more deliberately than previously thought to survey Antiquities systematically, rather than merely to study Antiquities for use in his paintings.[36]

• Very little is known about the history of restoration before the sixteenth century. The study of the restoration of ancient sculpture has suggested patterns across dispersed monuments.[37]

These connections, which could not be made in the original physical archives without many years of individual scholarship researching the source materials, indicate the sorts of questions that historians typically pursue. The Witt and Census information systems should be useful to architects, archaeologists, museum curators, art dealers, and historians of taste, scholarship, and culture in the broadest sense.

CONCLUSIONS

A data model has been described that attempts to address the requirements of an historical information retrieval system. This data model was designed to hold archival information the way it exists in an archive—sometimes fragmentary and sometimes contradictory. The Getty information system has been designed to replicate the archive in a computerized format where the researcher can browse, sift, and evaluate pieces of source information. But there is an important difference—the physical archive provides only a single mode of access in the order in which the documents have been arranged. Finding lists and catalogs, which are desired and used in historical research, provide a few additional "views" of the archive. The Getty information system adds complete random access from any recorded piece of information, or word within that piece of information, in the electronic archive, as well as a variety of formal cross-referencing between different types of information.

The Getty information system provides methods that support the extensive use of formal naming, which has been identified as a primary query mode of humanities researchers. The result is consistent forms of proper names, subject terminology, and other vocabularies for use when searching the electronic archive, while preserving the variations in naming as they exist in the source documents.

While the data model developed for the Getty information system appears to match well the research methods of historians, such a structured approach to information management requires intensive and time consuming information gathering and encoding. Because full text and multimedia databases with informal linking of data can be developed much more quickly and inexpensively with current computer technology, this method has recently become more popular than more structured methods. The quality and robustness of these different types of information systems need to be evaluated closely to determine which more fully meets the need for historical research.

END NOTE

This chapter was prepared originally as a paper for the second meeting of the International Society for Knowledge Organization (ISKO) held in Madras, India, August 26-28, 1992.[38]

The Witt and Census information systems are available at the host institutions in London and Rome, and at the Getty Art History Information Program in Santa Monica, California. For further information on access to the databases contact:

Dr. Catherine Gordon, Project Head, Witt Library Computer Index, Somerset House, Strand, London, WCR2 0RN, United Kingdom. Telephone: +44-71-873-2770/1. Fax: +44-71-873-2772.

Dr. Arnold Nesselrath, Project Head, Census of Antique Art and Architecture Known to the Renaissance, Bibliotheca Hertziana, Via Gregoriana 28, 00187 Roma, Italy. Telephone: +39-6-679-8325. Fax: +39-6-679-0740; or, Warburg Institute, Woburn Square, London, WC1H 0AB, United Kingdom. Telephone: +44-71-580-9663. Fax: +44-71-436-2852.

NOTES

1. Manfred Thaller, "The Historical Workstation Project." *Computers in the Humanities* 25, 1991, 154.

2. *The American Heritage Dictionary*, second college edition, Boston: Houghton Mifflin Co., 1985.

3. "The past and history float free of each other, they are ages and miles apart. For the same object on enquiry can be read differently by different discursive practices (landscape can be read/interpreted differently by geographers, sociologists, historians, artists, economists, etc.) whilst, internal to each, there are different interpretive readings over time and space; as far as history is concerned historiography shows this." Keith Jenkins, *Rethinking History*, London: Routledge, 1991, p. 5. Jenkins' brief book provides a clear discussion of the current thinking about history, and particularly the teaching of history, in a "modern" world. This paper assumes that the aim of historical study is research to discover evidence and construct an historically informed view, but not to discover or construct real "truth" or certainty.

4. Simon Schama, *Dead Certainties (Unwarranted Speculations)*, New York: Alfred A. Knopf, 1991, 322. This book considers two rather loosely related historical events: the death of General Wolfe during the siege of Quebec City, and the sensational 19th century murder of George Parkman, a wealthy Bostonian, by a professor at Harvard Medical College who was indebted to him.

5. Jenkins, 7. [see Note 3]

6. Helen R. Tibbo, "Information Systems and Services, and Technology for the Humanities." *Annual Review of Information Science and Technology* 26, 1991, 287-346.

7. Donald Owen Case, "Conceptual Organization and Retrieval of Text by Historians: The Role of Memory and Metaphor." *Journal of the American Society for Information Science* 42(9), 1991, 657-668.

8. Marilyn Schmitt, general editor, "Object, Image, Inquiry: the Art Historian at Work." Santa Monica, CA: Getty Art History Information Program, 1988. Tape recordings and transcriptions of Schmitt's interviews for this project as well as several interviews with architectural historians have been deposited in the archives of the Getty Center for the History of Art and the Humanities. These materials may be a unique source for further study of the research methods of humanists. Access to the archive can be sought by writing to the Getty Center Archives, 401 Wilshire Boulevard, Santa Monica, CA 90401.

9. Susan L. Siegfried and Deborah N. Wilde, "Scholars Go Online." *Art Documentation*, Fall 1990, 139-141. See also the following articles: Marcia J. Bates, Deborah N. Wilde, and Susan Siegfried, "An Analysis of Search Terminology Used by Humanities Scholars: The Getty Online Searching Project Report Number 1." *Library Quarterly 63*, January 1993, 1-39. Susan

Siegfried, Marcia J. Bates, and Deborah N. Wilde, "A Profile of End-User Searching Behavior by Humanities Scholars: The Getty Online Searching Project Report No. 2." *Journal of the American Society for Information Science* 44(5), 1993, 273-291.

10. Stephen E. Wiberley, Jr., "Names In Space and Time: The Indexing Vocabulary of the Humanities." *Library Quarterly* 58, January 1988, 1-28.

11. Stephen E. Wiberley, Jr. and William G. Jones, "Patterns of Information Seeking in the Humanities." *College and Research Libraries* 50, November 1989, 640-641.

12. "Usually the searching they did beforehand consisted of looking at a map or a telephone book to obtain repositories' addresses; if available, a published guide to the collection would be consulted." Wiberley and Jones, 1989, 643.

13. Marcia J. Bates, "The Design of Browsing and Berrypicking Techniques for the Online Search Interface." *Online Review* 13(5), 1989, 407-424.

14. Thaller, 1991, pp. 149-162. [See Note 1]

15. Thaller, pp. 155-159, describes the software model of the Max Planck "historical workstation." [See Note 1]

16. "The Electronic Pierce Consortium: A Network of Scholars and Technologies" was described in a session at the American Society for Information Science 54th Annual Meeting, October 31, 1991. The project aims to combine encoded electronic transcriptions of C.S. Pierce manuscripts with bit-mapped images of the pages, a bibliographic database, and hypertextual tools for analysis.

17. A database of medieval numeric data sponsored by the Research Library Group Program in Research Information Management. See: "The Medieval and Early Modern Data Bank." *The Research Libraries Group News* 12, January 1987, 8-10.

18. Described in a presentation at the Museum Computer Network meeting in Chicago in 1989.

19. Some portions of the discussion of the Witt Computer Index were first developed as part of a presentation on "Descriptors for Describing Pictures" given at the American Society for Information Science Annual Meeting technical session on "Image Classification Research: New Strategies and Techniques," October 20, 1989, by Joseph A. Busch and Cathy Whitehead.

20. The Witt Computer Index artist authority includes all the British, Canadian, Australian, and New Zealand artists whose works are represented in the Witt Library (approximately 18,000 artists). Catherine Gordon, "Dealing with Variable Truth: The Witt Computer Index." *Computers and the History of Art* 2(1) 1991, p. 27. The Witt has recently produced a computer-generated "checklist" of British artists derived from the authority file structures of the Computer Index database. The checklist contains artists' names, dates, and locations of birth and death, professional associations (including dates), notes on the extent of material in the Witt Library Collection, and variant names. *Checklist of British Artists in the Witt Library*, London: Witt Library, Courtauld Institute of Art, 1991.

21. Phyllis Pray Bober, "The Census of Antiquities Known to the Renaissance: Retrospective and Prospective." In: *Roma, centro ideale della cultura dell'Antico nei secoli XV e XVI Da Martino V al Sacco di Roma*, 1417-1527, Milano: Electo, 1989, 372.

22. Bober, p. 373.

23. Phyllis Pray Bober and Ruth Rubinstein, *Renaissance Artists and Antique Sculpture: A Handbook of Sources*, New York: Oxford, 1986.

24. Arnold Nesselrath, "Current Status of the Census of Antique Works of Art and Architecture Known to the Renaissance" [unpublished], 1990.

25. Carol McMichael Reese and Marilyn Schmitt [transcript of] "Interview with Rick K. Holt, Online Computer Systems, Inc.," Washington, DC, 1988, 15.

26. Joseph A. Busch, ["Conversation with Catherine Gordon, Witt Computer Index," Washington, DC, 1991].

27. The design was also informed by Holt's active participation in the planning and analysis for a number of other Getty projects, including the development of an editorial system for the Art and Architecture Thesaurus (AAT), and a data entry system for the International Repertory of the Literature of Art (RILA).

28. Although, as discussed in Wiberley and Schmitt, it may be acceptable that the research be mediated by an expert as is the case in most special collections such as archives. The most important criteria for usefulness by outsiders may be the ability to ask useful research questions and to be able to provide useful results. Thus the choice to focus the design on the database producers may have been a fully valid methodology.

29. See the essay "Iconography and Iconology: An Introduction to the Study of Renaissance Art." In: Erwin Panofsky, *Meaning and the Visual Arts*, London: Penguin, 1970. This essay, originally published in 1939, is the basis for modern iconographical analysis.

30. Deirdre C. Stam, "What About the Mona Lisa? Making Bibliographic Databases More Useful to Art Historians by Classifying Documents According to the Aspect of Art Object(s) Under Consideration." *Art Documentation*, Fall 1991, 129. Stam's typology is based on the assumption that an art object is essentially a kind of communication; as Stam argues, the Shannon-Weaver communication model could be analogously applied to it. See also Stam's earlier work on a communication model of art history in her dissertation, "The Information-Seeking Practices of Art Historians in Museums and Colleges in the United States, 1982-83," [New York], Columbia University, 1984.

31. Gordon 1991, p. 22. [see Note 20]

32. Reese, p.21. [see Note 25]

33. For example, Bober and Rubinstein, 1986 as cited above.

34. Except in the case where additional fields were added to hold the full-text transcription of primary source information, such as eyewitness accounts, original documents, or information inscribed on the objects and artifacts themselves.

35. This discovery was made by Catherine Gordon, but it has not yet been published.

36. Arnold Nesselrath, "Raphael's Archaeological Method." In: *Raffaello a Roma, il convegno del 1983*, Roma: Edizioni dell'Elefante, 1986.

37. Arnold Nesselrath, "The *Venus Belvedere*: An Episode in Restoration." *Warburg Journal* 50, 1987, 205-214.

38. Cognitive Paradigms in Knowledge Organization, Bangalore: Sarada Ranganathan Endowment for Library Science, 1992, 372-389.

BIBLIOGRAPHY

Bates, Marcia J. "The Design of Browsing and Berrypicking Techniques for the Online Search Interface." *Online Review* 13(5), 1989: 407-424.

Bober, Phyllis Pray. "The Census of Antiquities Known to the Renaissance: Retrospective and Prospective." In: *Roma, centro ideale della cultura dell'Antico nei secoli XV e XVI Da Martino V al Sacco di Roma, 1417-1527*, Milano: Electa, 1989, 372-381.

Bober, Phyllis Pray; Rubenstein, Ruth. *Renaissance Artists and Antique Sculpture, a Handbook of Sources*, New York: Oxford, 1986.

Case, Donald Owen. "Conceptual Organization and Retrieval of Text by Historians: the Role of Memory and Metaphor." *Journal of the American Society for Information Science* 42(9), 1991: 657-668.

Checklist of British Artists in the Witt Library, London: Witt Library, Courtauld Institute of Art, 1991.

Gordon, Catherine. "Dealing with Variable Truth: the Witt Computer Index." *Computers and the History of Art* 2(1), 1991: 21-27.

Gordon, Catherine. "An Introduction to ICONCLASS," [text of a paper presented at the MDA Meeting, 1988].

Gordon, Catherine. "The Witt Computer Index: Answering Unasked Questions." *Bulletin of the American Society for Information Science* 18(2), December/January 1992: 22-23. In special section "Art Information, Information Systems in Cultural Institutions," comp. and ed. by Joseph A. Busch.

Jenkins, Keith. *Rethinking History*, London: Routledge, 1991.

Nesselrath, Arnold. "Current Status of the Census of Antique Works of Art and Architecture Known to the Renaissance," [unpublished], 1990.

Nesselrath, Arnold. "Raphael's Archaeological Method." In: *Raffaello a Roma, il convegno del 1983*, Roma: Edizioni dell'Elefante, 1986.

Nesselrath, Arnold. "The *Venus Belvedere*: An Episode in Restoration," *Warburg Journal* 50, 1987: 205-214.

Panofsky, Erwin. *Meaning in the Visual Arts*, London: Penguin, 1970.

Reese, Carol McMichael; Schmitt, Marilyn. [transcript of] "Interview with Rick. K. Holt, Online Computer Systems, Inc.," Washington, D.C., 1988.

Schama, Simon. *Dead Certainties (Unwarranted Speculations)*, New York: Alfred A. Knopf, 1991.

Schmitt, Marilyn, general editor. *Object, Image, Inquiry, the Art Historian at Work*, Santa Monica, CA: Getty Art History Information Program, 1988.

Siegfried, Susan L.; Wilde, Deborah N. "Scholars go online." *Art Documentation*, Fall 1990: 139-141.

Stam, Deirdre C. *The Information-seeking Practices of Art Historians in Museums and Colleges in the United States, 1982-83*, New York: Columbia University, 1984.

Stam, Deirdre C. "What About the Mona Lisa? Making Bibliographic Databases More Useful to Art Historians by Classifying Documents According to the Aspect of Art Object(s) Under Consideration." *Art Documentation*, Fall 1991: 127-130.

Thaller, Manfred. "The Historical Workstation Project." *Computers in the Humanities* 25, 1991:149-162.

Tibbo, Helen R. "Information Systems and Services, and Technology for the Humanities." *Annual Review of Information Science and Technology* 26, 1991: 287-346.

Wiberley, Stephen E., Jr. "Names in space and time: the indexing vocabulary of the humanities." *Library Quarterly* 58, January 1988: 1-28.

Wiberley, Stephen E., Jr.; Jones, William G. "Patterns of Information Seeking in the Humanities." *College and Research Libraries* 50(6), November 1989: 638-645.

APPENDIX 1: Witt and Census Vocabulary and Authority Controlled Fields with Examples   Page 1

Record
Page

Field Name	Type of Field	Sample Values
Witt Index - Work of Art Record		
Basic Catalogue Details	single page	
Artist Name	single authority	INMAN, Henry; TWACHTMAN, John Henry; COLE, Thomas; EDDY, Don
Search Date	single date authority	1900-1980; 1778-1860; 1888; 1913
Subject Description	repeating subject authority	25K3; 46C131; 31AA15; 25G3
Last Known Ownership	single owner authority	Columbus, Museum of Art; Hotel Drouot; PONT, Henry Francis du; Kornfield & Klipstein
Type of Work	single fin ite vocabulary	mural; sculpture; painting; drawing
Mount	repeating page	
Image Shape	single extensible vocabulary	vertical rectanglar; horizontal rectangular; irregular; oval
Source of information	repeating link to work of art	COPLEY, John Singleton;1788;Hunting Scene
Component Parts	single page	
Description of Composite Work	single extensible vocabulary	companion pieces; pair; series; double sided
Order of Cataloguing	single extensible vocabulary	none perceived; narrative; numbered; spatial
Components	repeating link to work of art	COPLEY, John Singleton;1788;Hunting Scene
Part of	repeating link to work of art	COPLEY, John Singleton;1788;Hunting Scenes,companion pieces
Witt Index - Artist Name Record		
Name and School	single page	
National School	single extensible vocabulary	American; British; French; Italian
Personal name	repeating personal name authority	DUVENCK, Frank; PETO, John Frederick; LA FARGE, Thomas Cowperthwart; KRASNER, Lenore, Miss; FRANKENTHALER, Helen, Miss
Professional Association	repeating page	
Type of Association	single extensible vocabulary	member; associate; Royal Academician; National Academician
Date of association	single date authority	1779; 1886; 1898; 1792-1820
Institution	single institution name authority	Philadelphia, Art Clubs of; Pennsylvania, Academy of Fine Arts; Royal Academy of Arts; Society of Illustrators
Witt Index - Bibliography Record		
Bibliographic Definition	single page	
Type of Document	single extensible vocabulary	sale catalogue; exhibition catalogue; journal; book
Witt Index - Engraver Record		
Engraver Definition	single page	
Individual engraver	single personal name authority	HOPPER, Edward; DURAND, Asher Brown; CURRIER, Nathaniel; PEALE, Charles Wilson
Institutional engraver	single institution name authority	Bosqui; Sorony; Middleton, Strobridge & Co.; Bouve & Sharp
Witt Index - Exhibition Record		
Exhibition Organiser & Title	single page	
Organiser	repeating institution name authority	Indianapolis, Museum of Art; National Gallery of Art, Washington, DC; unspecified; Downtown Gallery, New York, NY; West, Benjamin; Copley, John Singleton
Exhibition Date & Location	repeating page	
Starting date	single date authority	1967:7:9; 1968:10:22; 1915; 1963:10
Ending date	single date authority	1975:11:16; 1982:8:1; 1964:5; 1956

APPENDIX 1: Witt and Census Vocabulary and Authority Controlled Fields with Examples   Page 2

Record
Page

Field Name	Type of Field	Sample Values
Venue	single institution name authority	Janis, Sidney, Gallery, New York, NY; National Gallery of Art, Washington, DC; Acquavella Galleries, New York, NY; unspecified
Bibliographic citation	single bibliography authority	National Portrait Gallery, Washington, DC., West & Am. Students, Wash.,DC; Acquavella Galleries, New York, NY, 1981, (May-June), 19th-20th c. Ptgs., New York, NY; Institute of Contemporary Arts, London, GTL., 1964, Screen Print Proj., London; Rhode Island, School of Design, 1972, To Look on Nature, Providence, RI
Location	single geography authority	Baltimore, MD; Washington, DC; Frankfurt am Main, DEU; Philadelphia, PA

Witt Index Geography Record
Geographic Definition — single page

Country	single extensible vocabulary	Italy; Great Britain; Netherlands; France
Variant Names	repeating geography authority	Atlanta, GA; Berlin, GER; New York, NY; Rome, ITA., see Roma, ITA; London, GTL; Washington, DC; Burlington House

Witt Index - Institutuion Name Record
Institution Definition — single page

Type of Institution	repeating extensible vocab	art dealer; auctioneer; museum; educational establishment; library; publisher; private collection
Preferred/Non-preferred	single finite vocab	yes; undetermined; no
Location	repeating geography authority	St. Louis, MO; San Francisco, CA; Cleveland, OH; Richmond, IN
Variant or preferred name	repeating institution name authority	National Academy of the Arts Design; American Society of Painters in Watercolor; Apollo Association; Herron, John, Museum of Art

Institution History — repeating page

Date	single date authority	1937; 1878-1937; early 20th c.; unspecified; late 19th c.; 1820-1910
Type of event	single extensible vocabulary	founded; in existence; exhibitions started; administered by; formerly named
Related person	repeating personal name authority	HAMMER, Armand, Dr.; STIEGLITZ, Alfred; MELLON, Andrew William; HALPERT, Samuel, Mrs., (Edith Gregor)
Related institution	repeating institution name authority	Jackson, Martha, Gallery, New York, NY; Harvard University; Smithsonian Institution; High Museum of Art
Related collection	repeating extensible vocabulary	Abbey, Edwin Austin Collection; Bruce, Ailsa Mellon Collection; Dale, Chester Collection; Trumbull Collection; Mellon, Andrew William Collection
Location	repeating geography authority	Albany, NY; Chapel Hill, NC; Dallas, TX; Paris, FRA

Witt Index - Personal Name Record
Personal Name — single page

Record
Page

Field Name	Type of Field	Sample Values
Variant Names	repeating personal name authority	POLLOCK, Jackson, Mrs., (Lee Krasner); CODMAN, Martha C., Miss; HOPPER, Edward, Mrs., (Josephine Verstille); ZSISSLY; ANNELLIO, Francis; PIKE, Alice, Miss
Biography	single page	
Date of birth	single date authority	1855; 1909; 1886; 1945
Location of birth	single geography authority	Oxford, OH; Portsmouth, NH; New York, NY; Khorkom Vari Halyotz Dzor, TUR; Montreal, CAN
Date of death	single date authority	1970; 1948; 1980; 1962; circa 1958
Location of death	single geography authority	Rockport, MA; at sea; New York, NY; France
Earliest known date	single date authority	1787; 20th c.; 1865; 19th c.; circa 1850
Last known date	single date authority	1977; 1975; 1983; 19th c.; 1928
Address	single geography authority	Upperville, VA; New York, NY; Philadelphia, PA; Paris, FRA
Relationships	repeating page	
Related person	single personal name authority	GORKY, Agnes Phillips; FEININGER, Theodore Lux; SLOAN, John; CALDER, Alexander Stirling
Type of relationship	single extensiible vocabulary	father; mother; son; daughter; employer; friend; pupil; teacher

Witt Index - Owner Record
Owner Definition   single page

Private owner	repeating personal name authority	ELLISON, Nancy & HELLMAN, Jerome; ELLISON, Robert S., Mrs., estate of; O'KEEFE, Georgia; PEALE, family
Institutional owner	repeating institution name authority	Fogg Art Museum; Busch-Reisinger Museum; Anderson-Meyer, Galerie, Paris, FRA; Harvard University

Witt Index - Sale Record
Auctioneer & Title   single page

Auctioneer	repeating institution name authority	Hotel Rameau; Hauswedell, Dr. Ernst; Parke-Benet; Anderson Gallery, New York, NY
Sale Date & Location	single page	
Starting date	single date authority	1938:10:18; 1915:1:12;1922:4:27; 1937:4:22
Ending date	single date authority	1927:1:18; 1967:2:8; 1969:3:20; 1949:3:31
Bibliographic citation	single bibliography authority	Parke-Benet, 1951:4:17, Pepsi-Cola Sale, New York, NY; American Art Association, 1916:2:21, Lambert Sale, New York, NY; Sotheby Parke-Benet, 1969:3:20, New York, NY; Keeler Art Galleries, 1922:5:4, De Forest Sale, New York, NY
Location	single geography authority	Philadelphia, PA; Boston, MA; New York, NY; Hamburg, DEU

Witt Index - Subject Index Record
Subject Index Description   single page

Index words	repeating extensible vocabulary	burial rites; commercial fishery; education; law & jurisprudence

Witt Index - Date Record
Definition   single page   *(not available for searching)*

APPENDIX 1: Witt and Census Vocabulary and Authority Controlled Fields with Examples   Page 4

Record
Page

Field Name	Type of Field	Sample Values
Type of date	single extensible vocabulary	year; month; range; day
Witt Index - Source Record		
Definition of Source Information	single page	
Type of work	single finite vocabulary	painting; drawing; construction; print
Type of source	single finite vocabulary	Witt; bibliographic; lettering; inscription
Bibliographic citation	single bibliography authority	
Advertisement	single finite vocabulary	yes
Medium & Size	single page	
Materials/Printing process	repeating extensible vocabulary	oil; aluminum paint; pencil; wax
Supports	repeating extensible vocabulary	canvas; board; paper; cloth
Techniques	repeating extensible vocabulary	airbrush; collage; pen; grisaille
Unit of measure	single extensible vocabulary	inches; centimeters; unspecified
Paper size	single extensible vocabulary	small folio; large folio; folio; elephant
Alternate Artists & Titles	single page	
Alternate attributions	repeating artist authority	STUART, Gilbert; HOPPNER, John; RAMAGE, John; SARGENT, John Singer; MARTIN, John; DANBY, Francis; GREENWOOD, John; AMERICAN SCHOOL 20thc.
Additional artists	repeating artist authority	AUDOBON, Victor Gifford; BARD, John; ALZIEUX, C.; ROSENBERG, Charles G.; TRUMBULL, John; COPLEY, John Singleton; TAIT, Arthur Fitzwilliam; MIGNOT, Louis Remy
Inscription(s)	repeating page	
Inscription type	repeating extensible vocabulary	signed; dated; initialled; dedicated
Inscription position	repeating extensible vocabulary	lower right; lower left; mount; specific
Print Details	single page	
Designer	repeating personal name authority	SARGENT, John Singer; WHISTLER, James Abbott McNeill; WEST, Benjamin; LEWITT, Sol
Engraver	repeating engraver authority	WOLF, Henry; WARD, William H.; WAY, Thomas, Robert; MOSES, Henry; NORMAND, fils
Publisher/Printer	repeating institution name authority	Moses, Henry; Boydell, John; Smith, John Raphael; West, Benjamin; Gemini, Los Angeles, CA
Place published	single geography authority	London, GTL; New York, NY; Los Angeles, CA
Date	single date authority	1831; 1779:11:4; 1970; 1972
State of print	single extensible vocabulary	proof; second; first; third
Lettering Details	repeating page	
Lettering type	repeating extensible vocabulary	artist identified; dated; sitter identified; inscribed
Lettering position	repeating extensible vocabulary	lower left; lower right; mount; lower centre
Exhibitions	repeating page	
Exhibition	single exhibition authority	Marborough Fine Art, London, GTL., 1963, R,B. Kitaj, London, GTL.; Heim Gallery, London, GTL, 1976, Am. Drawings, Hatch Coll., London, GTL; Boston, Museum of Fine Arts, 1983, A New World, Boston, MA; National Gallery of Art, Washington, DC., 1965, J.S. Copley, Washington, DC

APPENDIX 1: Witt and Census Vocabulary and Authority Controlled Fields with Examples   Page 5

<u>Record</u>
Page

Field Name	Type of Field	Sample Values
Lent by	single owner authority	Boston, Museum of Fine Arts; Tate Gallery; Royal Collection, Great Britain; SAATCHI, Doris Lockhart & Charles
Auction Sales	repeating page	
Sale	single sale authority	Sotheby Park-Benet, 1839:6:1, Benjamin West Sale, London, GTL; Charpentier, Jean, Galerie, Paris, FRA., 1935:6:7, Bornheim Sale, Paris, FRA; Parke-Benet, 1961:12:10, L.A. County Museum Benefit, Los Angeles, CA; Christie, Manson & Woods, 1976:12:3, London, GTL
Seller	single owner authority	WEST, Benjamin, Jr.; SCULL, Robert C., estate of; Baltimore Museum of Art; Prendergast, James, Library Association
Buyer	single owner authority	National Gallery, London, GTL; Aberdeen, Art Gallery; WEST, Benjamin; MARTIN, (buyer) I
Ownership	repeating page	
Owner	single owner authority	HATCH, John davis; White House, The; NETSCH, Walter Andrew, Mr. & Mrs.; Metropolitan Museum of Art
Date of event	single date authority	1968; 1976; 1925; 1818-1828; 1953-1957; 1930's; circa 1946-1953
Type of event	single extensible vocabulary	in possession of; acquired; disposed of
Acquired from	single owner authority	CANTOR, Joseph, Mr. & Mrs.; ALLSTON, Washington; HOOPER, Samuel, Mrs. & Alice, Miss; CABOT, Henry B., Mrs.
Type of transaction	single extensible vocabulary	gift; long term loan; bequest; diploma presentation; donation; commission
Last Provenance	single page	
Location(s)	repeating geography authority	London, CAN; Boston, MA; New York, NY; Massachusetts, USA; Italy; France
Ownership	repeating owner authority	Old Print Shop, New York, NY; Castelli, Leo, Gallery, New York, NY; BLOCH, Albert, Mrs.; SONNABEND, Ilena
At original location	single extensible vocabulary	no; yes
Date	single page	
Date	single date authority	1870's; 19th c.; 1960; 1910-1910
Related Works	repeating page	
Type of relationship	single extensible vocabulary	print after; copy; preliminary drawing; version; original
Work(s) of art	repeating link to work of art	
Work(s) of art not in Witt	repeating link to work of art	

APPENDIX 1: Witt and Census Vocabulary and Authority Controlled Fields with Examples   Page 6

<u>Record</u>
Page

Field Name	Type of Field	Sample Values
<u>Census - Monument Record</u>		
Description of Monument	single page	
Class	repeating extensible vocabulary	architecture; mosaic; sculpture
Type of building/object	repeating extensible vocabulary	temple; statue; sarcophagus
Original?	single extensible vocabulary	yes; no; unknown
Material/Medium	repeating extensible vocabulary	bronze; marble: luni/carrara; opus incertum
Present State	single page	
Present location	single location authority	Firenze, Uffizi; untraced; Roma, Regio IX (Circus Flaminius)
Present condition	repeating extensible vocabulary	damaged/fragment; intact; restored
Historic Origins	single page	
Artists/Creators	repeating person authority	Hadrian; Phidias; unknown
Date of creation	single date authority	1st cent. (BC); Trojan-Hadrian; 200-235
Style	single style authority	Etruscan; Augustan; Hellenistic
Renaissance attributions	repeating person authority	Phidias; Praxiteles; Agrippa, Marcus Vinsanius
Renaissance datings	repeating date authority	Augustus; Tiridates; Vespasian
Renaissance Provenance	repeating page	
Date	single date authority	16th cent.; 1550-1562; 1560, around
Renaissance location	single location authority	Firenze, Giardino de Pazzi; Roma, Campidoglio; Roma, Arco di Portugallo
Documentation	repeating link to document	El Escorial, inv. Codex Escurialensis 28-II-12; fol. 53 r| Firenze, Biblioteca Marucelliana, Ms. A. 79. I; fol. 7 v| Roma, BAV, vat. lat. 3616; fol. 31 r| Georgius de Negroponte 1507 (luzio 1886); Letter to Castiglione
First Renaissance Condition	single page	
Condition when first known to		
Renaissance	repeating extensible vocabulary	damaged/fragment; intact; restored
Reason known to Renaissance	repeating extensible vocabulary	always there; described in ancient literature; excavated
Documentation	repeating link to document	Vacca 1594 (Nibby 1820); no. 26 (p. 15)| Haarlem, Teyler's Stichting, inv. K III 21; r| Cambridge, Trinity College, Cambridge Sketchbook; fol. 5
Conservation History	repeating page	
Date	single date authority	Gregory III; 1444; post 1588:9:16-ante 1588:11:14
Action(s) taken	repeating extensible vocabulary	destroyed; restored; altered
Person(s) responsible	repeating person authority	Fontana, Domenico; Theodoric; Goths
Documentation	repeating link to document	Bartsch 1854-1870; vol. III, p. 45, no. 144| Aldroandi 1556; p. 206.E| London, BM, P&D, inv. 1946-7-13-639; r
Related Monuments & Documents	single page	
Part of	repeating link to monument	Vestal, statue; Circus Maximus, circus; Altar to Cybele and Attis Dedicated by L. Cornelius Scipio Oreitus, altar

APPENDIX 1: Witt and Census Vocabulary and Authority Controlled Fields with Examples   Page 7

Record
Page

Field Name	Type of Field	Sample Values
Component(s)	repeating link to monument	right flank, flank; Altar to Cybele and Attis Dedicated by L. Cornelius Scipio Oreitus, altar| Obeliscus Augusti, obelisk; Circus Maximus, circus| Altar of the Gods of Heaven and Earth, altar; Circus Maximus, circus
Replica of	repeating link to monument	Apollo, statue; Laocoon, group of statues; Aphrodite, statue
Replicas known to Renaissance	repeating link to monument	La Zingara, statue; Camillus, statue; Venus Santa Croce, statue
Parallel replicas of same original known to Renaissance	repeating link to monument	Camillus, statue; Mercury Farnese, statue; Amor Drawing Bow, statue
Documentation	repeating link to document	Aldroandi 1556; p. 206.E| Venezia, ASV, Domenico Grimani Inventory 1587; no. 4| Southeby's Sale, London, 1975:3:21; lot 5
Images	repeating image authority	BH0000860%; WI0000123D; BH00060472
Later Known Replicas	repeating page	
Location	single location authority	Basel, Antikenmuseum; Roma, Antiquario del Foro di Augusto; Muenchen, Glyptothek
Image(s)	repeating image authority	WI0000112B; BH00060393
Monument Bibliograhy	repeating page	
Citation	single bibliography authority	De Fine Licht 1968; Nush 1961-1967; Bober and Rubinstein 1986
Further Comments on Monument	repeating page	
Authorship of record	single extensible vocabulary	aan, me, Cassidy, Brendon Francis

Census - Renaissance Document Record

Identification/Reference	single page	
Document location	single location authority	Siena, Biblioteca Comunale; Roma, ASR; Berlin, SMPK, Kuperstichkab.
Publication	single bibliography authority	Aldroandi 1556; Agostini 1952; Albertini 1510 (Memoriale)
Type of documentary material	single extensible vocabulary	book of drawings, page(s): printed book, poem
Description of Document	single page	
Author(s) of document/Artist(s)	repeating person authority	Aldroandi, Ulisse; Da Caravaggio, Polidoro (attributed to); Anonymouse Pighianus
Date of document	single date authority	1510, around; post 1532:5:23-ante 1537:17:30
Type of written documentation	repeating extensible vocabulary	inventory, treatise, publication
Type of visual medium	repeating extensible vocabulary	drawings: pen, wash; engraving; statue: bronze
Method of representation	repeating extensible vocabulary	copy, elevation, modified version
Text	repeating page	
Context of text	repeating extensible vocabulary	location, nomination, comment: mythology
Location on document	repeating extensible vocabulary	top, bottom, left
Related Documents & Monuments	single page	
Monument(s)	repeating link to monument	Laocoon, group of statue; Unidentified Draped Female from the Belvedere Courtyard, statue; Mask with Phrygean Cap, mask
Part of	repeating link to document	Albertini 1510 (Opusculum)| Aldroandi 1556; p. 292.A - p. 295.B (House of Lorenzo Ridolfi)| Vico 1558

Record
Page

Field Name	Type of Field	Sample Values
Component(s)	repeating link to document	Aldroandi 1556; p. 122.E| Vico 1558; p. 40| Vasari 1568; parte III, p. 611
Prototype	repeating link to document	Firenze, Uffizi, inv. 1861 A; r| Arezzo, Casa Vasari, Salone; (place in order ?)
Copy/Copies	repeating link to document	Madrid, Prado, inv. 515; (place in order ?)| Firenze, Uffizi, inv. 1862 A; r
Parallel copy/copies of same prototype	repeating link to document	Firenze, Uffizi, inv. 1862 A; A 1862 r.A| Salzburg, Universitaetsbibliothek, ms. Ital. M III 40; fol. 27 r.A
Image(s)	repeating image authority	BH0000540; WI00020690; BH0005511I
Document Bibliography	repeating page	
Citation	single bibliography authority	Ashby 1904; Huelsen-Egger 1913-1916; Buddensig
Further comments on Document	repeating page	
Authorship of record	repeating extensible vocabulary	aan; me; Healy, Fiona
Census - Person Record		
Biographical Details	single page	
Date of birth	single date authority	post 50-ante 104; 1368; 1475:3:6
Date of death	single date authority	post 104-ante138:7:10; 1437:2:20; 1564:2:18
Place of origin	single location authority	Damascus; Firenze
Authorship of record	repeating extensible vocabulary	aan; me; Campbell, Ian
Census - Style Record		
Style Definition	single page	
Authorship of record	repeating extensible vocabulary	aan; me; Kleehsch, Ursula
Census - Location Record		
Topography	single page	
Part of	repeating location authority	Roma, Musei Vaticani, Cortile Ottagono; Braunschweg; Italy
Component(s)	repeating location authority	Roma, Museo Capitolino; Roma S. Maria dell'Aracoeli; Roma, Piazza di Campidoglio
Type of location	single extensible vocabeulary	courtyard, town, state
Authorship of record	repeating extensible vocabulary	aan, me
Census - Bibliography Record		
Bibliographic Details	single page	
Authorship of record	repeating extensible vocabulary	fgb; me; Stewering, Roswitha
Census - Image Record		
Image Data	single page	
Location of photograph	repeating extensible vocabulary	BH Raccolta di Disegni; WI Antiquities Census Sketchbooks
Source	single extensible vcocabulary	Courtauld Institute; Philadelphia, Rosenbach Foundation
Authorship of record	repeating extensible vocabulary	Riebesell, Christina; Healy, Fiona
Census - Date Record		
Date Details	single page	
Authorship of record	repeating extensible vocabulary	aan; me; Kleehsch, Ursula

Witt Index - Work of Art Record

Page 1: Basic Catalogue
Details — single page
Artist name — single authority
Short title — single text
Search date — single authority
Subject description — repeating authority
Last known ownership — single authority
Type of work — single finite vocab

Page 2: Mount — repeating page
Mount identification — single barcode

Witt microfiche number — single text
Witt accession date — single date
Courtauld negative number — single text
Other negative number — single text
Image shape — single exten vocab
Source of information — repeating link

Page 3: Component parts — single page
Description of composite work — single exten vocab
Order of cataloguing — single exten vocab
Notes — repeating text
Components — repeating link
Part of — repeating link

Witt Index - Artist Name Record

Page 1: Name and School — single page
Label — single unique text
National school — single exten vocab
Personal name — repeating authority
Note(s) — repeating text

Page 2: Professional
Association — repeating page
Type of association — single exten vocab
Date of association — single authority
Institution — single authority

Witt Index - Bibliography Record

Page 1: Bibliographic
Definition — single page
Label — single unique text
Type of document — single exten vocab
Full citation — repeating text
Note(s) — repeating text

Witt Index - Engraver Record

Page 1: Engraver Definition — single page
Label — single unique text
Individual engraver — single authority
Institutional engraver — single authority
Note(s) — repeating text

Witt Index - Exhibition Record

Page 1: Exhibition Organiser
& Title — single page
Label — single unique text
Organiser — repeating authority
Title — repeating text

Page 2: Exhibition Date &
Location — repeating page
Starting date — single authority
Ending date — single authority
Venue — single authority
Bibliographic citation — single authority
Location — single authority
Note(s) — repeating text

Witt Index Geography Record

Page 1: Geographic Definition — single page
Label — single unique text
Country — single exten vocab
State/Country/Province — single text
City/Town/Village — single text
Building — single text
Variant Names — repeating authority
Note(s) — repeating text

Witt Index - Institution Name Record

Page 1: Institution Definition	single page
Label	single unique text
Type of Institution	repeating exten vocab
Preferred/Non-preferred	single finite vocab
Location	repeating authority
Variant or preferred name	repeating authority
Note(s)	repeating text

Page 2: Institution History	repeating page
Date	single authority
Type of event	single exten vocab
Related person	repeating authority
Related institution	repeating authority
Related collection	repeating exten vocab
Location	repeating authority
Building	single text
Note(s)	repeating text

Witt Index - Personal Name Record

Page 1: Personal Name	single page
Label	single unique text
Surname/Anonymous	single text
First name	single text
Middle name(s)	single text
Prefix	single text
Suffix	single text
Variant Names	repeating authority
Note(s)	repeating text

Page 2: Biography	single page
Date of birth	single authority
Location of birth	single authority
Date of death	single authority
Location of death	single authority
Earliest known date	single authority
Last known date	single authority
Address	single authority

Page 3: Relationships	repeating page
Related person	single authority
Type of relationship	single exten vocab

Witt Index - Owner Record

Page 1: Owner Definition	single page
Label	single unique text
Private owner	repeating authority
Institutional owner	repeating authority
Note(s)	repeating text

Witt Index - Sale Record

Page 1: Auctioneer & Title	single page
Label	single unique text
Auctioneer	repeating authority
Title	repeating text

Page 2: Sale Date & Location	single page
Starting date	single authority
Ending date	single authority
Bibliographic citation	single authority
Location	single authority
Note(s)	repeating text

Witt Index - Subject Index Record

Page 1: Subject Index Description	single page
Label	single unique text
Index words	repeating exten vocab
Textual description	repeating text
Specific name identifiers	repeating text
Additional qualifiers	repeating text

Witt Index - Date Record

Page 1: Definition	single page
Label	single unique text
Date from	single date
Date to	single date
Type of date	single exten vocab
Note(s)	repeating text

Witt Index - Source Record

Page 1: Definition of Source
Information   single page
Type of work   single finite vocab
Type of source   single finite vocab
Bibliographic citation   single authority
Lot/Catalogue number   single text
Illustration number   single text
Page numbers   single text
Advertisement   single finite vocab

Page 2: Medium & Size   single page
Materials/Printing process   repeating exten vocab
Supports   repeating exten vocab
Techniques   repeating exten vocab
Dimensions   single text
Unit of measure   single exten vocab
Paper size   single exten vocab
Note(s)   repeating text

Page 3: Alternate Artists &
Titles   single page
Alternate attributions   repeating authority
Additional artists   repeating authority
Additional titles   repeating text
Standard reference   repeating text
Note(s)   repeating text

Page 4: Inscription(s)   repeating page
Inscription type   repeating exten vocab
Inscription position   repeating exten vocab
Transcription   repeating text
Note(s)   repeating text

Page 5: Print Details   single page
Designer   repeating authority
Engraver   repeating authority
Publisher/Printer   repeating authority
Place published   single authority
Date   single authority
State of print   single exten vocab
Note(s)   repeating text

Page 6: Lettering Details   repeating page
Lettering type   repeating exten vocab
Lettering position   repeating exten vocab
Transcription   repeating text
Lettering note(s)   repeating text

Page 7: Exhibitions   repeating page
Exhibition   single authority
Catalogue number   single text
Lent by   single authority
Note(s)   repeating text

Page 8: Auction Sales   repeating page
Sale   single authority
Lot number   single text
Seller   single authority
Buyer   single authority
Note(s)   repeating text

Page 9: Ownership   repeating page
Owner   single authority
Date of event   single authority
Type of event   single exten vocab
Acquired from   single authority
Type of transaction   single exten vocab
Accession/Inventory number   single text
Note(s)   repeating text

Page 10: Last Provenance   single page
Location(s)   repeating authority
Ownership   repeating authority
At original location   single exten vocab
Note(s)   repeating text

Page 11: Date   single page
Date   single authority
Note(s)   repeating text

Page 12: Related Works   repeating page
Type of relationship   single exten vocab
Work(s) of art   repeating link
Work(s) of art not in Witt   repeating link

Census - Monument Record

Page 1: Description of
Monument   single page
Class   repeating exten vocab
Type of building/object   repeating exten vocab
Original?   single exten vocab
Number of identical versions   single number
Material/Medium   repeating exten vocab
Descriptive details   repeating text
Descriptive coding   repeating text
Name(s)/Title(s)/ Identifier(s)   repeating text

Page 2: Present State   single page
Present location   single authority
Inventory numbers/Subdivision   repeating text
Present condition   repeating exten vocab

Page 3: Historic Origins   single page
Artists/Creators   repeating authority
Date of creation   single authority
Style   single authority
Renaissance attributions   repeating authority
Renaissance datings   repeating authority

Page 4: Renaissance
 Provenance   repeating page
Date   single authority
Renaissance location   single authority
Inventory number(s)/
 Subdivision   repeating text
Documentation   repeating link

Page 5: First Renaissance
 Condition   single page
Condition when first known to
 Renaissance   repeating exten vocab
Reason known to Renaissance   repeating exten vocab
Documentation   repeating link

Page 6: Conservation History   repeating page
Date   single authority
Action(s) taken   repeating exten vocab
Person(s) responsible   repeating authority
Documentation   repeating link
Note(s)   repeating text

Page 7: Related Monuments &
Documents   single page
Part of   repeating link
Component(s)   repeating link
Replica of   repeating link
Replicas known to Renaissance   repeating link
Parallel replicas of same
 original known to Renaissance   repeating link
Documentation   repeating link
Images   repeating link

Page 8: Later Known Replicas   repeating page
Location   single authority
Inventory number(s)/
 Subdivision   repeating text
Note(s)   repeating text
Image(s)   repeating authority

Page 9: Monument
Bibliography   repeating page
Citation   single authority
Page reference   single text
Illustration   single text

Page 10: Further Comments on
 Monument   repeating page
Comments   repeating text
Authorship of record   single exten vocab

Census - Renaissance Document Record

Page 1: Identification/
Reference	single page
Document location	single authority
Inventory number(s)/Name(s)	
of document	repeating text
Publication	single authority
Place in order/Reference	single text
Type of documentary material	single exten vocab

Page 2: Description of
Document	single page
Author(s) of document/Artist(s)	repeating authority
Date of document	single authority
Type of written documentation	repeating exten vocab
Type of visual medium	repeating exten vocab
Method of representation	repeating exten vocab
Dimensions of document	single text

Page 3: Text
	repeating page
Context of text	repeating exten vocab
Transcription of text	repeating text
Location on document	repeating exten vocab
Note(s)	repeating text

Page 4: Related Documents &
Monuments	single page
Monument(s)	repeating link
Part of	repeating link
Component(s)	repeating link
Prototype	repeating link
Copy/Copies	repeating link
Parallel copy/copies of same	
prototype	repeating link
Image(s)	repeating authority

Page 5: Document
Bibliography	repeating page
Citation	single authority
Page reference	single text
Illustration	single text

Page 6: Further comments on
document	repeating page
Comment(s)	repeating text
Authorship of record	repeating exten vocab

Census - Person Record

Page 1: Biographical Details
	single page
Name	single unique text
Date of birth	single authority
Date of death	single authority
Place of origin	single authority
Variant name(s)	repeating text
Note(s)	repeating text
Authorship of record	repeating exten vocab

Census - Style Record

Page 1: Style Definition
	single page
Style category	single unique text
Date from	single date
Date to	single date
Note(s)	repeating text
Authorship of record	repeating exten vocab

Census - Location Record

Page 1: Topography
	single page
Place	single unique text
Part of	repeating authority
Component(s)	repeating authority
Type of location	single exten vocab
Name(s) of buildings/	
institution	repeating text
Name(s) of collections	repeating text
Authorship of record	repeating exten vocab

Census - Bibliography Record

Page 1: Bibliographic Details
	single page
Abbreviated citation	single unique text
Full citation	repeating text
Note(s)	repeating text
Authorship of record	repeating exten vocab

Census - Image Record

Page 1: Image Data	single page
Bar code number	single bar code
Description of photograph	repeating text
Location of photograph	repeating exten vocab
Source	single exten vocab
Negative identification	single text
Videodisc number	single number
Videodisc frame	single number
Authorship of record	repeating exten vocab

Census - Date Record

Page 1: Date Details	single page
Date label	single unique text
Date from	single date
Date to	single date
Note(s)	repeating text
Authorship of record	repeating exten vocab

Chapter 3

ANALYZING ART OBJECTS
FOR AN IMAGE DATABASE

Lois F. Lunin

INTRODUCTION

Today's information systems can offer us images and text. Now, in addition to reading about art objects, we can see—if not the original—an image of the original. This exciting advance in technology raises some interesting questions: Now that we can see images does that mean we no longer need to describe an object? But if we do need descriptive terms, how will they relate to the image? Besides using words, is there another way to describe an image, for example, by selecting components of the image itself to represent it? And, since each presents advantages and disadvantages, should the image be in analog or digital form?

This chapter examines some of the facets to consider in planning an art information system that incorporates images of fiber art. Although focused on a specific form of art, the discussion applies to any object regardless of discipline.

DEFINITIONS AND EXAMPLES

The term *fiber art* can be vague, confusing, even unknown to many people. Yet the use of fibers as an art medium has a long history. "The use of fibrous materials as a medium for art works is not new; woven, knitted, printed, and otherwise treated materials have long appeared in the history of mankind." [Henning, 1977] Traditionally, however, fiber objects appeared as functional works. Fiber art, sometimes called *art fabric*, was introduced after World War II to describe new art developments in textiles. This chapter deals only with fiber art developments since 1945 and the challenges in describing the physical appearance and condition, composition, content, design, and intent of the art for inclusion in databases.

A definition of fiber art that satisfies everyone is as difficult to achieve as a definition of multimedia. And fiber art itself represents multimedia or mixed media in that it can include fiber—natural or synthetic—and numerous other materials such as glass, clay, metal, paper, and plastic.

Preceding the term fiber art as a classification were the words *fabric* and *textile*. While fabric might be accepted as the generic term for all fibrous constructions, the art fabric can be thought of as an artist's individual construction. [Constantine and Larsen, 1980]

The art fabric can be woven on the loom or free of the loom, or it can be produced by a variety of other techniques, for example, knotting, twining, and interlacing. A distinguishing characteristic is that art fabric is most often conceived and created by one artist whose efforts, talent, and passion fuse with his/her technical abilities and materials. In common with most artists, the artist who works with fiber possesses the same artistic education and background, uses technological advances including computers, and enjoys experimentation and manipulation of materials that have stimulated new concepts in all types of twentieth century art.

EXAMPLES OF FIBER ART

To talk about fiber art without showing examples is to invite a serious injustice to the art. The works must be seen to be understood and valued. Yet even pictures do not adequately illustrate a work's dimension, shape, density, volume, color, texture, material, shadows, and movement, although they impart at least a sense of what we describe as fiber art. Thus, the four photographs reproduced here (pp. 3-6) only suggest a bit of the form and variety found in fiber art. The first three works are owned by the Smithsonian's Renwick Gallery of the National Museum of American Art.

Each work differs from the other in material, technique, shape, texture, density, size, and expression. Because these pictures present only the frontal view of the works, one cannot peer through, feel, or see on their surfaces the effects of increasing or diminishing light. Some of these limitations can be ameliorated in an image information system that permits an image to be manipulated to provide greater contrast with the background, and to be enlarged, rotated, or otherwise enhanced.

HISTORY OF FIBER ART

The terms fiber art or art fabric came into use to describe the work of the artist-craftsman following World War II. [Constantine & Larsen, 1973; Brite & Stamsta, 1986; and others] Artists began to reevaluate the loom—long the basic equipment in producing fabric—as an expressive tool. Weavers began to see that they could bind fibers into nonfunctional forms with the validity of a work of art. [Nordness, 1970] Artists in the United States and Europe explored the qualities of fabric or linear elements of linen, sisal, cotton, etc. to develop both wall hangings and freestanding works; of two or three dimensions, flat or volumetric; many stories high or miniature; nonobjective or figurative; representational or fantasy; rough-textured or gauze-like; environmental that one might walk through or canopy-like to protect and entice. They also created fiber structures using methods not requiring a loom, such as wrapping, casting, collaging, binding, and coiling. (continued, page 62)

Figure 1, "Coil Series II—A Celebration," by Claire Zeisler, was made in 1978. The piece is 65 inches high by 34 inches in diameter; it is freestanding, constructed of natural hemp and wool. An elegant, imposing, dramatic, lyrical structure, the three-dimensional work begins with a symmetrical arch. The yarns cascade gracefully to the base where, wrapped over a core of intricately intermeshed coils, they spill onto the floor.

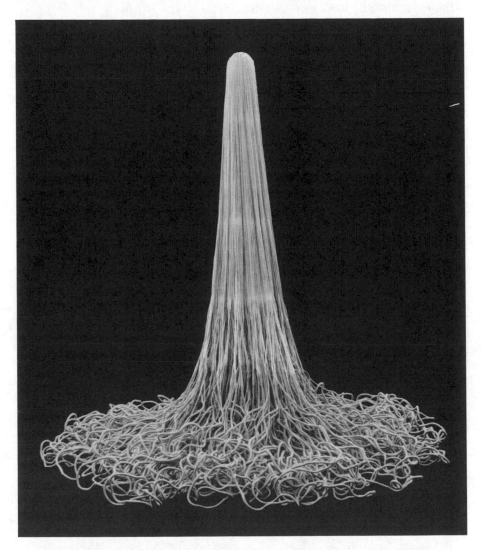

Figure 1. "Coil Series II—A Celebration" by Claire Zeisler. Photo courtesy of the Smithsonian Institution's Renwick Gallery of the National Museum of American Art. Photo credit: Bruce Miller.

Figure 2, "Mourning Station #11" was made by Dominic DiMare in 1988. Constructed of wood, horsehair, feathers, and clay, the work is 44 inches by 22 inches by 22 inches. Delicate, exquisitely conceived and crafted, DiMare's forms clearly distinguish his works from his contemporaries. "Although his intricately constructed structures often seem related to religious icons as well as to ritual objects of the Eskimo and Native Americans, they are clearly not derivative of them." ["Ten Works Acquired in 1992 for Renwick Gallery's Collection."]

Figure 2. "Mourning Station #11" by Dominic DiMare. Photo courtesy of the Smithsonian Institution's Renwick Gallery of the National Museum of American Art. Photo credit: Bruce Miller.

Figure 3 shows a wall hanging made by Mariska Karasz in 1950 titled "Skeins." Karasz was one of the pioneers in the development of fiber art in the U.S. in the 1940s and 1950s. Embroidered handsomely, the work is made of linen, cotton, and wood and measures 50 inches by 54 inches. Steeped in traditions of embroidery that tend toward fixed patterns with little deviations, Karasz revolutionized stitchery by using those traditional techniques in a way that brought a personal and unique quality into her work. ["Ten Works Acquired in 1992 for Renwick Gallery's Collection."]

Figure 3. "Skeins" by Mariska Karasz. Photo courtesy of the Smithsonian Institution's Renwick Gallery of the National Museum of American Art. Photo credit: Bruce Miller.

Figure 4, "Untitled #81," shows a work by John McQueen done in 1979. It is wickerwork made of red osier, walnut bark, and ash; it measures 18 inches by 10 inches by 24 inches. "A basket with a comet's tail is constructed of sticks whose forked ends are united by weaving. The tail implies direction and movement; the container results from the forced fusion of divergent linear paths into a circular enclosure." [*Rossbach and Halper*, 1992].

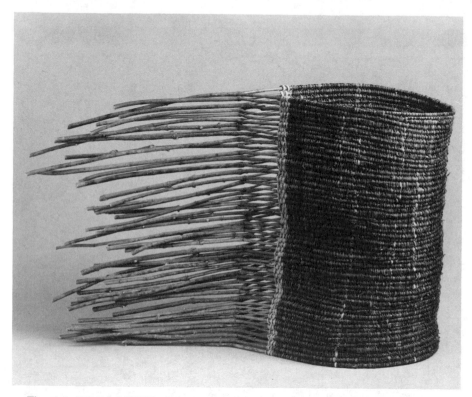

Figure 4. "Untitled #81" by John McQueen. Photo courtesy of the Smithsonian Institution's Renwick Gallery of the National Museum of American Art. Photo credit: Edward Owen.

HISTORY OF FIBER ART *(continued)*

The post World War II artist-craftsman movement led to the development of the studio artist and revolutionized the creative concept of the object. Lenore Tawney, a weaver, was one of the first. Beginning in the late 1950s, she explored the creation of three-dimensional forms with "constructions evoking the power and spatial relationships of sculpture." [Nordness, op. cit.] The opening of her exhibition at the Staten Island Museum in 1961 marked the first major exhibit of American art fabrics.

Recognizing the art of fiber beginning with their first and continuing through their tenth exhibitions, the Biennales Internationales de la Tapisserie in Lausanne, Switzerland showed a shift from tapestries (group productions by painters and artisans) to fiber works conceived and executed by artists; from two-dimensional mural textiles to three-dimensional works; from works with an aesthetic emphasis on imagery to those that relied upon the textural or structural qualities of textiles. Other changes reflecting the current aesthetic concerns were also appearing in other contemporary art forms.

"Fiber R/Evolution," a landmark exhibition in 1986 developed by the Milwaukee Art Museum, separated the displayed items into a revolution component displaying works by the creators of the new movement, and an evolution portion showing works by the artists whose efforts often grew from or were stimulated by the earlier work. [Brite & Stamsta, op. cit.] Today the importance of fiber art grows ever stronger as evidenced by an increasing number of exhibitions; exhibition catalogs, and scholarly and popular books; serious collectors; and formal organizations founded specifically to promote and advance creative techniques.

Yet art abstracting and indexing services reflect the newness of the fiber art field by the paucity of their descriptive terminology. Indexing and abstracting services only recently added the term *textiles*. [Shaw, 1990] *Art Index* first used the word *fiber* in volume 19, November 1971-October 1972. *RILA, The International Repertory of the Literature of Art*, appears to have used the term *fiberwork* beginning with the 1980-84 issue; and *ART Bibliographies Modern* first used the term in 1988.

IMAGES AND IMAGE INFORMATION

Images are perceived variously according to each viewer's background, training, visual perception and neural processing, politics, and many other factors. Thus, each of us viewing an object might expound a different description and interpretation. In viewing fiber art—which often is textural, nonobjective, or nonrepresentational—it is difficult to realize an object's structural components and the effect it manifests when displayed as, for example, sitting on a shelf, hanging overhead, inviting us to walk through, or allowing us to step on. Thus, at this juncture of time and technology, some description in text form seems to be required to describe facets that a flat image on a screen cannot communicate.

Yet the image is necessary because, for some of the same reasons stated above concerning individual perception, background, and politics, words alone cannot give the whole picture. Also, many users require both the intellectual and the graphic elements to understand an image. Studies such as Rorvig's [1987] demonstrate the necessity of considering both the amount of description for images as well as the relation between the physical data image and its pointer surrogate. Thus, verbal descriptions of fiber works, at least at the present state of the art, need to be considered in databases of images.

MATERIALS

Contemporary artists have access to vast resources of different fiber types, natural as well as synthetic. Artists today combine fibrous materials with thread, clay, paper, wood, metal, paint, glass, and other materials. As noted in the descriptions of the four art works pictured earlier, their creators used hemp, wool, wood, horsehair, feathers, clay, red osier, walnut bark, ash, linen, and cotton. Materials and objects culled from the descriptions of fiber works illustrated in just two publications—*Fiber R/Evolution* [Brite & Stamsta, 1986] and *The Art Fabric Mainstream* [Constantine & Larsen, 1980]—numbered 162 [see Lunin, 1990, Table 1 for complete list]. This list shows that almost everything and anything can be included within a work of fiber art, and begins to illustrate some of the complexities involved in using words alone to describe fiber art.

TECHNIQUES

Many processes are used in producing fiber work, some dating to prehistoric times. Weaving, for example, is one of the earliest techniques, but many other methods, such as twining, knotting, wrapping, sewing, and felting, even preceded weaving. The Jacquard loom, invented in 1780 in France, was an ancestor of today's computer-assisted loom. The Jacquard loom was controlled by punched cards, a forerunner of the Hollerith cards used so extensively by first-generation electronic computers.

No less a figure than Leonardo da Vinci may have given first thought to mechanized weaving in describing the technique as ". . . second only to the printing press in importance; no less useful in its practical application; a lucrative, beautiful and subtle invention." [quoted in DeGraw, 1972] In likening the loom to the printing press, Leonardo surely qualifies as an early information professional whose foresight predated SIG/AH with its focus on information technology in the arts and humanities.

Techniques for creating fiber art vary almost as much as materials used. Most of the techniques were known in antiquity, but some new procedures have been developed—an amalgam of creative minds and hands with new technology. From the two books cited earlier a list of 42 techniques were identified, e.g., couching, interlacing, looping, shibori, and soumak. [See Lunin, 1990, Table 2 for complete list.] A more detailed listing of techniques can be found in the ARTSearch Techniques Table: Field Descriptions and Valid Field Values [1988]. And as stated in the description of the table's "type" field: "Just as there can be several techniques used to create an object, there can be several types within each of these techniques. There is no limit on the number of techniques or types within techniques that can exist for one object."

VOCABULARIES

It is encouraging to note that a vocabulary is evolving to describe the appearance and meaning of fiber art. The terms come from many sources and many different players. For example, in submitting work to an exhibition, an artist often is required to write a statement of intent, and new terms might emanate from those descriptions. The jurors sometimes use another vocabulary in judging the submitted work. Curators, art critics, writers and editors of art books, art educators, and gallery directors may use even somewhat different terms. The variety of these vocabularies illustrates still another area of complexity. The same two sources cited previously provide 212 terms, among them *abstract, brooding, gossamer, dynamic, intricate, mysterious, pendulous, rich, tactile, sensual,* and *wispy.* [See Lunin, 1990, Table 3 for complete list.] However, as pointed out by Brandford [1990] these terms are not at all specific or unique to fiber.

DESIGN

The artists working with fiber as a medium usually are motivated by an aesthetic rather than a utilitarian or functional need. In other words, art takes precedence over function: the work is a wall hanging rather than a tablecloth; a container is a sculpture rather than a basket for holding fruit or nuts. To some viewers, however, the wall hanging might look like a tablecloth and the container might be purchased for use. Still, it is the aesthetic aspects that dominate in the object's creation. The elements of line, color, texture, shape and form, and principles of rhythm, unity, balance, and emphasis—the basic elements and principles of design—take precedence. [Rutherford, 1989]

CREATORS AND USERS OF FIBER ART VOCABULARIES

Those who describe fiber art use a rich language. The registrar or collection manager is responsible for the transportation, packing, storage, and handling of all objects brought into the museum for exhibitions, works lent to other institutions, pending acquisitions, and the recording and documenting of those works. [Ricciardelli, 1987] The registrar uses terms to indicate such aspects as museum number, artist, title, date, medium, dimensions, and condition of the work.

The curator is concerned with conceptualizing, planning, and selecting works for exhibitions, and for research in the collection. In considering a fiber piece for exhibit or purchase, for instance, the curator wants to see the object first hand because few pictures can give the feeling of the texture, the luminosity, and the impact that a large piece can produce when viewed both at a distance and up close.

The art historian uses a vast network of resources to explore, reaffirm, or negate previous assertions about a work and the culture in which it was created. Fiber art is created throughout the world and in greatly varying cultures. The materials, design symbolism, sex of the artist, and culture of the period all concern the art historian who, like the curator but with a different objective, wants access to many photos for comparison of details of similar work and perhaps any rituals associated with its creation.

The art conservator focuses on the preservation of the object. Because fiber artists use many man-made materials—plastic garbage bags, electronic wire, etc.—conservators must continue to learn about the aging of these materials. Will they yellow, crack, disintegrate, attract insects, absorb moisture and swell, thus placing a strain on other fibers in the piece? The conservator is the doctor of textiles, prescribing the treatment, stabilization, restoration, and mounting for installation. To assist conservators the Getty Conservation Information Network facilitates the retrieval and exchange of information about conservation and restoration of cultural property. The network includes three online databases—bibliographic, materials, and suppliers.

There are other users, too: iconographers, crime detection officers, art educators, students, collectors, insurers, gallery directors, art administrators, critics, artists, art librarians, editors, publishers, restorers, suppliers, writers, and the public. Each brings a particular focus of interest to the search or to the writing—and each uses a somewhat different vocabulary. [See Lunin, 1990, Table 4 for a more detailed list.]

THE LONGITUDINAL RECORD

Like other pieces of art, a record accompanies a work of fiber art. The record begins when the piece is created; it includes information on the creator(s); full demographic data; education and accomplishments of the artist; where the work was exhibited; reproduction of the work in a catalog, newspaper article, book, and more. The health of the work is also important: it may not be strong enough for moving to other museums or to be included in traveling exhibitions. And if it has a record of repairs, the nature of those repairs must be listed. The record must be open-ended and continued throughout the life of the object—and perhaps even beyond if the object itself has been destroyed, as in a fire, or lost, as in a theft.

DATABASE RECORD

Today's technology allows for at least four types of database records with information about images. The first is the conventional database with its many text fields to describe the image. Here the description refers to the image but does not include the image itself. A second is the field database that enters information about the object in a field but refers to its storage in another location. The third is a database of just the images. Some multimedia databases now include images of photographs and paintings that are located by identifiers or keywords describing some of the characteristics of the image or the artist and title of the work. A fourth kind—Binary Large Object Database (BLOB)—includes the pixels of the digitized image in a relational database. Size is no restriction: here the pixels are tucked or pushed into the record. [Lunin, 1992]

In time we expect that database records will be redesigned if only to meet the challenges now raised by multimedia, of which the images are one component. For now, however, the most compelling component of a database describing fiber art is the terminology used within the record fields. Some of the relevant fields are listed below [see Lunin, 1990, Table 5 for a more detailed listing].

- Artist
- Title of Work
- Execution Date, Year Produced
- Media, Material, Fiber Content
- Technique
- Structure(s)
- Type of Equipment Used
- Theme, Subject
- Style, Period
- Color(s)
- Dyes Used
- Texture
- Decoration, Surface Embellishment
- Design Symbolism
- Size, Dimension
- Owner
- Provenance
- Reproduction (Photos)
- Exhibition History
- Accession Number
- Location
- Appraised Value
- Insured Value
- Keywords

Scott [1988] reports that the catalog database for sculpture at the National Gallery of Art would use at least 300 tags "breaking down materials, techniques, iconography, and stylistic factors in detail." It is not too far-fetched to expect that a description of fiber art might require even more details.

TERMS AND FACETS

Museum specialists and others are increasingly aware that more facets of art need to be addressed than those now appearing in the abstracting and indexing publications and in-house catalogs. In the 1960s museum and computer system designers looked to computers for help in the organization, storage, search, and retrieval of art information. More than a decade ago the *Art & Architecture Thesaurus* (*AAT*) was begun. The initial intent was to provide catalogers with terminology to describe objects, documents about objects, and object and document surrogates—the AAT encompasses more than 30,000 terms in 36 separate hierarchies. Those hierarchies describe physical attributes, styles and periods, agents, activities, and materials and objects, but do not include subject description.

Description needs for fiber art include the structural and aesthetic characteristics of the art fabric as an art form, rather than just its materials and methods. Just as a pottery bowl begins with clay or a gold art necklace begins with the metal, however, it is with the material and method that the art fabric begins.

The field of art object cataloging is just beginning to recognize the inadequacy of language as a recording medium for describing a work of art. [Stam, 1989] But a simple description of the physical object is not enough. As Stam has stated, it is "the significance of the piece—a concept representing a perceiver's judgment—based on any one of several criteria." The groups of data which she indicated that need to be provided include "objective data about the work; subjective or interpretive data; style; evaluation; and even more today—signs, signification, and social context."

Stam sees that a redefinition of the problem is due not just to more sophisticated understanding of the art data but also to several recent technological advances: the hard disk; improved communication modes; relational databases; and software with flexibility in field definition and manipulation.

IMAGE DATABASES

As noted earlier, although words can conjure up a mental image, these same words can produce a different interpretation in each reader or listener. To clarify the description, an image surrogate offers visual facets of information about the art—its design, structure, strength of lines, and artistic subtleties. [Lunin, 1987]

Some fiber image information systems do exist. One is the Helen Allen Textile Collection located at the University of Wisconsin. This collection contains about 12,000 textiles, costumes, and related objects ranging from pre-Columbian and Coptic fragments to contemporary fiber art. An interactive laser videodisc computer system called ARTSearch was developed to meet both the intellectual and viewing access needs.

The University of Maryland Historic Textile Database, established in 1986, contains information about 10,000 coverlets. Implemented on a personal computer with a sophisticated data management program, the system can handle the massive amounts of data necessary for research on historic textiles. The purpose is to be able to search for and compare motifs in the same and different geographic areas and to study the popularity, uniqueness, origin of motifs, and migration patterns. The system uses PictureWare and an image capture board. While coverlets are not fiber art as defined for this chapter, the database serves as an example because it contains images of textiles. [Anderson, 1991]

SO WHERE ARE WE IN TERMS OF TEXT?

Fiber art still needs more specific as well as broader nomenclature and terminology. For some users it would be helpful if the abstracting and indexing services included additional descriptive terminology, enabling the user to conduct a search with more specificity. Also, the need for a more complex vocabulary should be explored when the indexing of images is considered.

As with most databases, fiber art databases will attract varied users with varied needs. Some users will need much description, perhaps even full text. Art historians, for example, prefer to work with a large quantity of information and, as Jost [1986/87] explains, they "will forego the comforts of standardized and integrated systems which offer little [limited] information in favor of a large quantity of less-structured data, even if it means working with several different databases of various listings and thesauri."

There are problems in coding fiber images. Indexing them for access to satisfy a number of interests is still an uncertain proposition because they can be perceived from many different perspectives. [Bearman, 1988, 1989] And even with much information available concerning the work, it is difficult to describe the concept and other important aspects with just a few index terms or a classification. As for representing a specific work with words, filling out the long form inherent in supplying detailed information as proposed for fiber art records is time-consuming and thus expensive.

In addition, real and basic differences separate the documentation of a bibliographic work and an art object. Documentation of a bibliographic work acts as a pointer to the literature in the book or article. Documentation of an art object is, as Barnett [1988] explains, a description of an otherwise mute subject.

In discussions with a group of people interested in fiber art, I asked whether each one would like images included in a database. Their responses, together with opinions expressed in the literature, indicate almost universal agreement on the desirability of having both the image and textual information.

ANALOG OR DIGITAL?

Whether the image should be analog or digital remains as another topic of discussion and controversy. Videodisc (analog) form seemed suitable for most needs. Videodisc stores the image in black and white or color; the image can be identified and located in the file of like objects (by virtue of an index or classification) and the disc can be distributed to many locations for use.

In contrast, digital systems store the image data in pixel (picture element) form, thus offering many different possibilities. The image can be manipulated—rotated, zoomed up and down, given scale, placed in an architectural environment, compared with other images fairly easily, and even studied microscopically to see its structure more clearly. But digital recording and transmission of images, even with data compression, require vast amounts of storage space; and a significant amount of transfer time is needed to bring the image to the screen.

With the growing popularity of multimedia and the evolution of new technologies, systems will be able to handle digital images—or analog images—more easily and less expensively at a desktop PC or workstation. For example, with the development of new compression and decompression technologies such as DVI (Digital Video Interactive), an all-digital multimedia environment offers several new opportunities for users of art databases.

HYPERTEXT, HYPERMEDIA

Many users of art databases would find it useful if records could be linked in both hypertext and hypermedia. Yet, as Bearman [1989] wrote, the technical and conceptual limitations of our approaches to multimedia humanities knowledge bases make it unlikely that we will see any universal products in our lifetimes. He offers the hope, however, that we might still construct some exciting, if limited, multimedia bases for particular types of users.

CHALLENGES

Some of the problems and challenges in describing fiber art are included above. These can be extended to the general problem of terminology and description for any visual object. Although pattern recognition is being developed to identify technique, color, and shape, words still are needed to yield information on history, cultural interpretations, composition, and intent of some of the elements included in the fiber art. Perhaps in some future time we humans can use our several senses to see an object, to feel its real (original) composition, to smell it (for sometimes odor is a valid test to distinguish real from imitation), to taste it (as far as I know, this exists at present only for holographic chocolate [Lunin, 1991]), and to hear it (if the object produces sound, as some fiber art does waving in a breeze). So far, however, even with multimedia, it is unlikely that we can soon go beyond recognition of image, sound, and text.

On the linguistic side, associated trails to information that lead to further understanding of a work can be helpful. This approach might involve techniques such as hypertext and hypermedia. Additionally, the comments of the scholarly users of the system might supplement a record with a kind of running citation index.

As in medicine, a longitudinal health record that begins with the information about the parents (the creators) and ends only with the death and autopsy report might contain the life record of the object. Certainly, this information would interest art historians and other scholars who learn from the past to benefit the present and future.

Another component that may still be needed in designing the optimal database record for art are some basic user studies as well as more knowledgeable analysis of the information itself with follow-up studies to determine how this information is sought, captured, and used. Future studies of the use of multimedia and the design of the secondary databases to locate the varied contents of multimedia may give us richer data about how we recognize and select image information.

THE REAL SIGNIFICANCE OF FIBER ART

As with any art, it is not the intellectual aspects of fiber work that are important but rather how it was created and the response it generates—excitement, an interest in the composition or structure, a strongly negative reaction. A political work of fiber art might offer a new way of looking at the world—and stir controversy even at the highest levels of government. How to document such intangibles requires keen perception, carefully selected words, rapport with the art itself, and, as perceptions and responses change, a running record to determine whether the work is exhibited or held in storage.

Whether the image is described or stored in analog or digital form is not really important. What matters is the artist's creative pursuit of new concepts, investigating a variety of materials and techniques to carry the message, pushing at the boundaries. At some point an artist experiences a kind of breakthrough when fresh possibilities appear—an original art form that has no precedent. The real challenge in information science is how to identify and describe that event, that magic moment, for one's contemporaries and for future generations.

SUMMARY

The field of fiber art and the art fabric continues its intense activity, exploration, and experimentation. Descriptions of that art should do no less than mirror its energy and dedication. This chapter suggests some kinds of terms needed by people working in aspects of fiber art. These needs can be extended to all forms of art and, indeed, to all objects. As many of us who have worked on thesauri and design of information systems know, an emerging and evolving discipline requires flexibility in its terminology to accommodate new techniques, materials, and uses. Images will help to broaden an understanding of those objects, but words still provide the contextual and intentional information.

While continually developing information technology offers the opportunity to do almost incredible things with images, the technology cannot supply the interpretation offered by carefully selected verbal descriptions. Thus, the situation is not either/or, but rather accretion—adding the sense of vision to the understanding that language alone provides.

DEDICATION

This chapter is dedicated to Jack Lenor Larsen whose vision, creativity, and gracious generosity continue to enrich his many contributions to the field of fiber and art.

NOTES

Anderson, C. S. "A User's Applications of Imaging Techniques: The University of Maryland Historic Textile Database." In: Lynch, C. A. and Lunin, L. F. (eds): Perspectives on Imaging: Advanced

Applications. *Journal of the American Society for Information Science* 42(8): 597-599, 1991.

"ARTSearch Techniques Table: Field Descriptions and Valid Field Values." Madison, WI: School of Family Resources and Consumer Sciences, University of Wisconsin at Madison, 1988.

Barnett, P. J. "An Art Information System: From Integration to Interpretation." *Library Trends* 37(2): 194-205, 1988.

Bearman, D. "Considerations in the Design of Art Scholarly Databases." *Library Trends* 37(2): 206-219, 1988.

Bearman, D. "Implications of Interactive Digital Media for Visual Collections." *Visual Resources* 5: 311-323, 1989.

Brandford, J. S. [personal communication], 1990.

Brite, J. F. and Stamsta, J. *Fiber R/Evolution.* Milwaukee, WI: Milwaukee Art Museum, 1986.

Constantine, M.; Larsen, J. L. *The Art Fabric: Mainstream.* New York: Van Nostrand Reinhold, 1980.

Constantine, M.; Larsen, J. L. *Beyond Craft: The Art Fabric.* Tokyo: Kodansha International, 1973.

DeGraw, I. *Fibre Structures: An Exhibition of Contemporary Textiles.* Denver, CO: Denver Art Museum. [a catalog of an exhibition that took place May 9 to June 18, 1972]

Henning, E. B. *Fibreworks.* Cleveland, OH: Cleveland Museum of American Art, 1977.

Jost, K. "Quantity is Quality: The Best Database is Worthless Without Data." *AICARC Bulletin of the Archives and Documentation Centers for Modern and Contemporary Art* 2(1): 49-50, 1986-87.

Lunin, L. F. "Electronic Image Information." In Williams, M. E., (ed.) *Annual Review of Information Science and Technology,* 24: 179-224. Amsterdam, The Netherlands: Elsevier, 1987.

Lunin, L. F. "The Descriptive Challenges of Fiber Art." *Library Trends* 38(4): 697-716, Spring, 1990.

Lunin, L. F. "Compound Documents, Multimedia, and Edible Information." *Information Today,* 42-43, April, 1991.

Lunin, L. F. "Image Overview '92." *Proceedings of the Thirteenth National Online Meeting,* New York City, May 5-7, 1992. Medford, NJ: Learned Information, 1992, pp. 195-202.

Nordness, L. *Objects: USA.* New York: Viking Press, 1970.

Ricciardelli, E. "Collections Management Within the Museum of Modern Art." *Spectra* 14(4): 10-11, 1987.

Rorvig, M. E.; Turner, C. H.; Moncada, J. "The NASA Image Collection Visual Thesaurus." Paper presented at the American Society for Information Science 17th Mid-Year Meeting. Ann Arbor, MI, 1988.

Rossbach, E.; Halper, V. *John McQueen: The Language of Containment.* Renwick Contemporary American Craft Series. [catalog of an exhibition held in 1992 at the Smithsonian Institution's Renwick Gallery of the National Museum of American Art, 1992.]

Rutherford, E. J. "Decisions: Exploring Line." [Abstract] In *Abstracts—7th Annual Conference on Textiles, June 23-25,* p. 32. College Park, MD: University of Maryland, Department of Textiles and Consumer Economics.

Scott, D. W. "Museum Data Bank Research Report: The Yogi and the Registrar." *Library Trends* 37(2): 130-141, 1988.

Shaw, C. [personal communication], 1990.

Stam, D. C. "The Quest for a Code, or a Brief History of the Computerized Cataloging of Art Objects." *Art Documentation* 8(1): 7-15, 1989.

"Ten Works Acquired in 1992 for Renwick Gallery's Collection." *Renwick Quarterly,* p. 2, Jun-Aug, 1992.

Photos were supplied courtesy of Michael Monroe, Curator-in-Charge, Smithsonian Institution's Renwick Gallery, National Museum of American Art.

INDEXING OF HYPERMEDIA

INTRODUCTION

Raya Fidel

Indexing is most commonly associated with the notion of linear text. Hypermedia introduces a two-fold innovation: information is limited neither to text nor to linear representation. Does hypermedia require, then, a new method of indexing? Is there a difference between the "old" methods of subject access and those that are suitable for hypermedia?

Much research and development are required before conclusive answers can be found to those questions. Initial discussions suggest, however, that much of the traditional indexing methodology is valid and important for successful retrieval from hypermedia systems.

Gary Marchionini sets the stage for the discussion. His chapter, "Designing Hypertexts: Start with an Index," provides a brief overview of hypertext systems and applications, and shows how authors of hypertexts can apply principles and techniques of indexing to resolve problems related to node and link management. He discusses the relationships between indexing and linking as viewed by various communities engaged in hypertext research and development. Issues addressed include what gets indexed (the nodes, the links); the extent to which the end user should be involved in indexing; and when should indexing be done (while the document is being created or after it is completed). Marchionini recommends: "Do the index BEFORE you start writing." Experience with this approach to developing a controlled vocabulary list to guide hypertext design provided insight to, and facilitated the development of, a structured procedure. This procedure for hypertext authoring includes a number of steps that are common in indexing and thesaurus construction.

Nancy Mulvany brings another example of the applicability of indexing principles and techniques to multimedia in "Online Help Systems: A Multimedia Indexing Opportunity." She observes that online help systems that are multimedia projects perform poorly for two reasons: a lack of index structure and awkward software tools. When computer users turn to help systems, they usually look for an immediate answer to a problem; having to browse through several screens would likely irritate most users. Yet, multimedia help screens are usually heavily layered. In addition, finding information is difficult because of limited search capabilities (e.g., no Boolean searching of the index) and poor design of the index structure. A sample list of principles developed for text indexes illustrates their usefulness in the design of online indexes. Nonetheless, indexers can successfully apply such principles to online indexes only when suitable software tools are available.

Peter Liebscher supports the idea that hypermedia producers should look at the work of indexers for solutions to some of their problems. His chapter, "Hypertext and Indexing," identifies many similarities between creating indexes and creating hypertexts. Viewing an index as an abbreviated description of a document or as an organizing structure for a document reveals that even linear documents contain several virtual, embedded indexes (e.g., table of contents, author's outline), some of which are dynamic. While index terms alone are dissimilar to links in that the terms do not express relations, an index entry *acts* like a link, expressing the relations "represents." Both hypertext authors and indexers determine relationships between concepts in their documents and then select appropriate symbols, such as index terms or icons. Documents and hypertexts differ only in their physical structural possibilities. A document has only one physical structure, and its conceptual structure can alter only through a surrogate. In contrast, the structure of a hypertext, along with its conceptual structure, can be easily altered. While this difference is fundamental to the user, the intellectual effort required for authoring is the same.

Dagobert Soergel further promotes the idea that the similarities between indexing hypermedia and indexing text are more pronounced than the differences. His chapter, "Information Structure Management," looks at database systems, expert systems, information retrieval systems, and hypermedia systems. These systems share an essential unity and a unified framework can improve their design. The systems differ along several dimensions, but usually only as a matter of degree. A new definition of a system structure generalizes search and interface operators that are applicable to any type of searching. The structure consists of objects (or entities) and relationships (or links) that make statements about objects. To enhance searching and inference, the structure includes neighborhoods (sets of objects with their relationships), especially offspring neighborhoods and ancestor neighborhoods, and connections (named chains of links). Assuming that navigation and query-based searching represent the same basic process, a discussion about approaches to searching addresses search and inference procedures, such as spreading activation (with Boolean and weighted searching as special cases); hierarchical inheritance; and structure matching. Indexing is the process of establishing the structure.

These chapters illustrate that indexing of hypermedia as well as authoring such projects are similar to indexing linear text documents. At the same time, examining retrieval from hypermedia systems suggests a new, generalized approach to searching systems and to indexing.

Chapter **4**

DESIGNING HYPERTEXTS:
START WITH AN INDEX

Gary Marchionini

INTRODUCTION

Hypertext is text in electronic form that permits or encourages readers and authors to work in nonlinear sequences. Hypertexts are designed for storage and manipulation on computer systems—they consist of units of information called "nodes" connected by relationships called "links." Links may be specified in advance and/or interactively. Those that are specified in advance suggest a type of indexing.

Hypertexts are user-directed information sources and thus similar to libraries that support self-directed information seeking and learning. Information professionals have therefore taken great interest in the development of hypertext technology and the research and design community has begun to seek the help of librarians, indexers, and other information professionals in their development efforts. This chapter provides a brief overview of hypertext systems and applications, discusses problems related to node and link management that may benefit from indexing practice, and describes an authoring strategy that has proven useful in developing group-authored hyperdocuments.

Several hypertext systems or "shells" exist that allow creation of the applications that we call *hypertexts* or *hyperdocuments* or *databases*, depending on the speaker. Some of the larger scale systems include Notecards from Xerox [Halasz, 1988] and KMS, which has been used in a number of academic, corporate, and government environments [Akscyn, McCracken, & Yoder, 1988]. Intermedia was certainly the most widely known large-scale hypermedia system dedicated to educational applications [Yankelovich, Meyrowitz, Haan & Drucker, 1988]. The personal computer environment includes systems or environments such as HyperCard, SuperCard, Plus, and a number of others for the Macintosh platform; and at least two dozen systems for the MS-DOS Platform, including Hyperties, Toolbook, HyperPad, and Guide, which runs in both the Macintosh and the MS-DOS operating systems [see Horton, 1991 for a listing of various systems]. In fact, most database, word processing, and even some statistical packages claim to have some "hyper" this or that features. There is a growing literature devoted to hypertext system development [see Akscyn, 1991 for a comprehensive collection of papers distributed in hypertext form], but the focus here is on the development of hyperdocuments.

Hypertext systems are being used for a broad range of applications [see Frisse & Cousins, 1992 for an overview of applications and models]. In system documentation, systems such as Concordia [Walker, 1988] and Document [Girill, 1991] have been used for large-scale computer documentation projects. A host of applications are available in education including Landow's work with Intermedia [Landow, 1989] and the Perseus Project [Crane, 1988]. See Jonassen & Mandl, 1990 for a collection of educational applications. Many libraries have developed hyperdocuments to assist users [e.g., see the St. Paul Public Library HyperCard Stack and Vaccaro & Valauskas, 1989] and museums have begun to use hyperdocuments to augment exhibits [e.g., Brethauer, Plaisant, Potter & Shneiderman, 1989]. In entertainment, a number of commercial products have emerged for a variety of audiences, for example, the Manhole (ActiVision, 1988) for young children who have not learned to read and novels by Joyce [Joyce, 1990] and others. In the writing environment, work at the University of North Carolina with the Writing Environment [Smith, Weiss & Ferguson, 1987] has yielded theoretical and practical benefits for individual and group authoring. The best example and description of how hypertext technology affects the written word is provided by Bolter's *Writing Space*, a text available in both paper and electronic forms [Bolter, 1991]. In information retrieval, there are a number of examples, including The Virtual Notebook at the University of Houston Medical School [Burger, Meyer, Jung & Long, 1991] that integrates objects from external sources (e.g., text from the Internet) as well as internal services within the University, and the researcher's personal notes. The Dynamic Medical Handbook [Frisse, 1988] represents an excellent example of a reference work that integrates information retrieval techniques such as relevance feedback with hypertext linking capabilities. Most CD-ROM reference databases and full-text encyclopedias or collections also offer some hypertext features—if only the ability to jump between highlighted query terms resulting from full-text searches.

Because there has been so much "hype" and so many systems available, we see a variety of hyperdocuments emerging and are beginning at least to understand the problems involved in authoring and using hypertext. Nielsen [Nielsen, 1989] conducted a meta-analysis of hypertext applications from a usability perspective and determined that performance effects due to specific design are overshadowed by user and task variables. Thus, general principles for optimizing usability are unlikely to be found. Nonetheless, focused studies of user performance with specific systems have improved design and informed our understanding of human-computer interaction. For example, studies of the SuperBook system yielded dramatic improvements in subsequent revisions of that system [Egan, Remde, Gomez, Landauer, Eberhardt & Lochbaum, 1989]. Studies of the Perseus hypermedia corpus are beginning to yield insights into how students and instructors are able to learn and teach with hypermedia

[Marchionini & Crane, in press; Neuman, 1991], and have provided formative feedback on revisions of the corpus and on the infrastructural requirements necessary to use it. Bernstein [Bernstein, Bolter, Joyce & Mylonas, 1991] has developed statistical methods for analyzing the structural forms of hypertexts and applied them to fiction and nonfiction hypertexts, and Botafogo [Botafogo & Shneiderman, 1991] has developed formal methods for describing the structure of hypertexts. These methods have good potential for assessing hypertexts and for assisting in their continued development, although many problems associated with developing and using hyperdocuments remain.

One activity that remains highly problematic is the development and management of links. Designers have not thought of indexing and managing links in an organized and structured way—a lot of linking has been quite ad hoc. Some of the principles and techniques associated with traditional indexing must certainly be useful in developing hyperdocuments.

INDEXING AND HYPERTEXT

Indexing involves selective access—we want to provide people with "ways" to get to a particular thing. These ways, or paths, involve pointers or signposts that users apply to accomplish their access goals—these pointers are links, the essence of hypertext [Conklin, 1987]. Links facilitate selective access. At any given time a user may have one or many links available and can follow them or not, depending on decisions made in context—a form of selective access. Thus, following links in a hyperdocument is like using footnotes or an index in a paper document. Just as a book may contain both a back-of-the-book index and "see also" references or footnotes in the content, hypertext systems support both types of links (although much more attention has been given to the "see also" type). The buttons, hot words, or other link anchors in most hypertexts are like "see alsos" within a hyperdocument and can be thought of as "local indexes" because they are so embedded in context and pertinent to that context. To understand why indexing and linking have been loosely coupled in the past, it is necessary to distinguish some characteristics of hypertexts and to highlight some differences in perspective in the various communities engaged in hypertext research and development.

Scope of the Hyperdocument

It makes a great deal of difference whether we are concerned with within-document indexing or linking, or across-document linking. A hypertext that is specific to one subject is quite different from a collection of hypertexts, just as a large distinction exists between indexing a single book and indexing a library of books. For example, consider the granularity issue: Should the smallest unit of indexing be a letter, a tri-graph, a word, a paragraph, a region (if it happens to be an image in a hypermedia document), a volume of text, or an entire library or network in the case of the virtual, connected library of the world?

Original or Retrospective Authoring

It also makes a great deal of difference whether one is creating new text or aggregating and organizing existing text. The former activity is termed *original* and the latter *retrospective* authoring.[1] In original authoring, the author typically creates the links along with the nodes (one hopes) according to some general plan. Retrospective authoring is a new type of activity that combines the roles of a book editor/collator and an indexer. Akscyn's work in creating the *Hypertext Compendium* is an excellent example of how this new type of activity works. The "editor/linker" must know not only the machinations of the system for purposes of importing and manipulating text and creating links, but must create a conceptual scheme for organizing discrete units of text and decide where and how to link those units. People in these roles can likely learn from those who have developed indexes and indexing languages.

Explicit and Implicit Links

DeRose [1989] distinguishes in his taxonomy between *extensional* and *intensional* links. Extensional links are direct and unpredictable and intensional links are predictable from the form and structure of the document. Extensional links are explicit, exemplified by iconic buttons and highlighted words, and have received the bulk of the attention in hypertext developments. In some respects, explicit links are like a book's "see also" footnote references and implicit links are like a back-of-the-book index. Explicit links are hardwired—they are direct connections between nodes or between intermediate anchors (Akscyn called these intermediate anchors "stepping stones" in the *Hypertext Compendium*)—and thus difficult to manage in volatile or open hypertexts that undergo frequent revisions or updates.

The simplest example of intensional links are word lookups and these have been called implicit links. Mylonas [Mylonas & Heath, 1990] focused on implicit links in designing the Perseus corpus, which is viewed as a library or resource for scholars, teachers, and students. Implicit links require specific initiation by the user, e.g., a pull-down menu or initiation of string-search at any given time. Implicit links put more decision-making authority and responsibility in the hands of users, who must initiate linking rather than simply deciding whether to follow a suggested (explicit) link. From an authoring perspective, it is much easier to support simple implicit links since data independence can be maintained between nodes. At one level, concordance and statistical analysis routines are all that is needed to create such "indexes" to support implicit links. As any indexer knows, however, an index requires the extraction of concepts not just words, and substantial intellectual effort is required to create menus or indexes that allow users to use implicit links across essential concepts in the database.

---

1. See [Marchionini, Liebscher & Lin, 1991] for a treatment of these distinctions. In that work, these were termed "a priori" and "post hoc" authoring.

Static or Open Hypertexts

Although all hypertexts exist in electronic form, they may or may not be dynamically changeable. Some systems, for example, distinguish between authoring and using modes and do not allow end-users to make changes. Many hypertexts are static, not allowing the user to change or add information to the document, regardless of what the system will support. Some allow users to mark sections, make separate notes or add nodes. Other hypertexts are fully open or volatile [Bernstein, et al., 1991], not only permitting, but encouraging the user to fully interact with an evolving document. It is unclear what role indexing can play in volatile hypertexts, where change to the content is not only ongoing but much of what a user "extracts" from the document is personal interpretation. It may be argued that continual indexing is even more critical in such environments, but it is difficult to identify how such indexing can be made operational without automated tools that can facilitate implicit links.

Community Perspectives

Hypertext is both a display technology and a retrieval technology. One problem is that hypertext systems designers typically are computer scientists who think of hypertext shells as display technologies, while users—information scientists—think of hypertext systems as information retrieval technologies. Those who use existing shells to develop specific hyperdocuments must reap the benefits of what both communities offer. This is beginning to occur as evidence increases of cooperation and understanding across those communities.[2]

What Gets Indexed?

These distinctions between types of hypertexts illustrate the richness of the technology and several ways that indexing knowledge can be helpful to hyperdocument designers. To illustrate some of the challenge and potential, consider what actually gets indexed. (Note that indexing is used here quite broadly to include any organizational finding strategies that may be available to eventual end users or only to authors.)

Everyone agrees that the nodes should be indexed—we need pointers to content and in selective cases these pointers are actual links in the hypertext. Several of the chapters in this volume discuss this issue [e.g., see Chapters 6 and 7], and the example discussed later in this chapter is concerned with indexing of nodes.

We should also think about ways to index the links themselves. How do we begin to organize links? If we are allowed to have link types, e.g., pro and con links that clearly indicate to the user whether the reference point will support or refute the referent context, it is a simple step to develop lists of link types. In such a case, users may want to look at all of the links, pull out the links that are supportive, or if there

---

2. For example, a panel at Hypertext '89 included system designers, indexers, and information scientists [Bernstein, 1989].

are 150 supporting links and only 3 refuting links, conclude that the hypertext does not represent a very critical treatment of the topic.

Temporal links may also be created and these must be managed. For example, an author may want to create a link from a node to one that does not yet exist. A facility is needed to save it, put it on hold, put it on a stack, or use some kind of demon to monitor progress until some later time when it is appropriate to form the link. How do we manage numerous virtual links in a large-scale project?

Another link management problem involves computed links—those that initiate a process. For example, applying a computed link may initiate a routine that examines the user's path history and determines which of several target nodes to reveal. Are all possible targets associated with the link? Is the computational method indexed too? How can the user search such polyhierarchies?

Another type of indexing involves the structure of documents themselves. Documents that include markup codes can provide users access to structural (e.g., section heading, quotations) or typographic details. Some text retrieval packages allow users to conduct searches on Standard Generalized Markup Language (SGML) codes within a document. Such tools are useful to authors and users alike, whether they are working at syntactic or rhetorical levels. These capabilities also offer opportunities for customizable publishing on demand. Tools to analyze the structure of hyper-documents have been proposed [e.g., Bernstein, et al., 1991; Botafogo & Shneiderman, 1991] to reveal gross structural details. Bernstein's link apprentice tool creates a link plot (a cross-product map of nodes that indicates the presence of links between all pairs of nodes) for a document and thus supports additional linking and editing. Botafogo's work treats the hyperdocument as an acyclic graph and computes metrics for the respective distances between all nodes. The resulting values for each node provide good indications for attributes such as centrality and suggest which aggregates can be broken off and treated as supernodes in a larger organizational view of the document.

In addition to the hyperdocument itself, a number of secondary artifacts must be managed. Paths through the database, notes, or bookmarks that develop as users interact with the document must eventually be organized. HyperCard's "recent" feature displays the previous 42 cards and is useful for a single user using a hypertext in a single session, but what happens in shared environments or in very large environments used over time? Individual users can perhaps come up with some scheme, e.g., saving various paths or histories and giving them times and dates and appropriate names.

In fact, some systems, such as Perseus, provide a path feature and a path index to assist in managing and manipulating them separately from working with the corpus. Peter Evans, at the University of Maryland, has developed a very nice generic path type of tool that is an add-on to HyperCard stacks and allows the user to capture time and card locations and make notes.[3] It has been particularly useful for testing and debugging purposes.

As more histories or paths are developed, how should they be indexed? Personal collections may not be such a large problem, but as these become shared, either through the Internet or through a local area network, where many people are using the same system, there must be some kind of control over, at the very least, simple naming of objects and files.

INDEXING AND AUTHORING: AN EXAMPLE

Consider the problem of authoring a hyperdocument. We are in serious need of basic tools to assist in managing links. We have been forced to use several different tools and steps to take ASCII files, concatenate them, run them through a program that creates a concordance, and take the word lists to develop a vocabulary list for making links [Marchionini, Liebscher & Lin, 1991]. To do this in creating the *Hypertext Compendium*, Rob Akscyn [1991] used the KMS system. Although he is a pioneer in hypertext technology and familiar with most of the literature, he used KMS tools to make sure that when he created his topical index, the things that he knew were going to be main topics would be augmented by actual terms that occurred many times in the corpus. Thus, the tool serves as an augmentation of the human editor/linker's intellect. We need many more tools like this to help us in making links, especially in the retrospective kind of authoring. Tools for automatic indexing [e.g., Salton, 1989, Chapter 9] will surely be useful in assisting an author in identifying and verifying possible links in a hyperdocument.

Although tools will certainly be developed to assist in the authoring process and experience with hand-crafted and automatic indexing will undoubtedly influence their development, there must be an overall intellectual framework that shapes any logical expression of ideas. I suggest that a simple form of index can serve as the framework for beginning the authoring process. That is, do the index BEFORE you start writing. This perspective is based on experience with coordinating group-authored hyperdocuments designed and implemented by students in my classes over the last several years.

Our approach is to begin with a general topic (past topics have included electronic publishing, library automation, and electronic networking) and identify key facets that serve as conceptual strands for the hyperdocument. [See Marchionini, Liebscher & Lin, 1991 for an explication of an early example]. Once the key facets have been identified, a controlled vocabulary is developed and students write definitions and/or articles for those terms that, in turn, reference other terms in the collection. The Hyperties system used for these projects allows synonyms to be specified, so some flexibility is allowed on terminology. A small group of students acts as editors and linkers and is responsible for editing the articles for style and consistency, importing the ASCII files supplied by the authors and forming links between articles. In the 1992 Spring semester, this method had evolved to the point where the class developed a hyperdocument consisting of 168 articles (nodes) and 1231 links. En-

---

3. The Interlogger suite of HyperCard scripts is available from the University of Maryland Technology Liaison Office.

titled "Netguide," the database contains over 300 kilobytes of text and has been distributed freely to people throughout the country.

Procedure

The object was to assemble a resource for individuals interested in learning about networking. The topic was one of several suggested in a class discussion. Once the topic was agreed upon, seven main strands or facets of the topic were identified. These included five main topic areas (basic concepts, network applications and services, descriptions of specific networks, standards, and issues); a set of procedures for getting started with networking at the University of Maryland; and a guide to materials and resources (see Figure 1 for a graphic overview of this toplevel structure). For each of the strands, one student volunteered to serve as group leader and one or two other students joined the group. A set of five students agreed to serve as editor/linkers for the document. (These five eventually divided the tasks of editing for grammar and style from the tasks of importing and linking the various files that served as nodes.) Since students were studying networking and hypertext as two of the computer applications in the course, they were prepared to participate in a brainstorming session that generated 450 words or phrases related to the five subtopics. This list was alphabetized and shared, and the editor/linkers created a controlled vocabulary consisting of about 150 main terms and various synonyms. These main terms became the basis for "articles" (nodes) that were written by the students in the various groups.

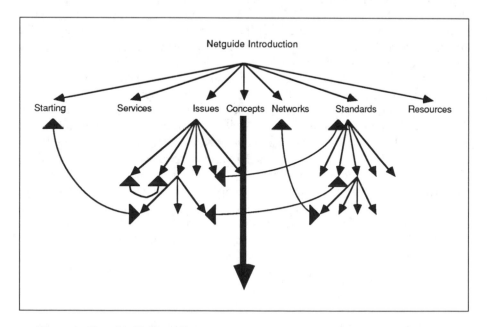

Figure 1. Netguide Toplevel Structure

The development of the article titles and mapping of these controlled terms to the other terms in the overall list was an important task for the editor/linkers. The links to these nodes were done by individual authors but links and nodes in this system are inextricably interwoven because links point to the top of an article (node). Thus, the editor/linkers constrained both nodes and links by creating the controlled vocabulary for the overall hyperdocument. Note that there is no implied sequence in this set of nodes; unlike a topic outline, it serves as an unordered specification of concepts. The decision to use a hierarchical structure for the entry and top-level nodes was based on experience with authoring and with results of studies of how people read linear texts and hypertexts.

Articles could range from a simple definition to small essays requiring multiple screens. The articles were composed using whatever word processor the author preferred and saved as separate ASCII files according to a naming scheme based on the controlled vocabulary. Within each article, use of other controlled vocabulary terms were identified by tildes (The Hyperties 2.3 system used for this project used tildes to identify the start and finish of each link anchor). Synonyms and commentary were saved in a separate file and turned in with the articles as suggestions to the editor/linkers. For example, if an author used the term "sign-on" in an article and "logon" was the controlled term, they could note this use and the editor/linker could make the link from the "sign-on" occurrence (link anchor) in the text to the "logon" article.

The editor/linkers edited the files for content and style and returned them to the authors who made changes and resubmitted the final files. It should be noted that electronic mail and file transfer were used extensively to facilitate exchange and "meetings" among author and editor/linker teams. The editor/linkers imported the ASCII files into Hyperties 2.3 and examined the first draft for gaps and overlaps. A few additional articles were requested and the hypertext was distributed for final comment and editing by the entire group. At the end of the fifteen-week semester, the hyperdocument was completed and shared with others beyond the class.

Results and Analysis

To examine such a corpus (to say nothing of grading it!), required traversing the hypertext in a variety of ways. Since it is infeasible to try every possible path, strategies for examining the document's overall structure were developed. For each of the 168 articles, the number of links coming in and going out was determined (Hyperties supports this minimal type of analysis). For example, the article "bulletin board system" has seven "synonyms" (including several morphological variants): BBSS, BBS, Bulletin Board, Bulletin Board Service, Bulletin Board System, Bulletin Board Systems, and Bulletin Boards. There are twelve links into this article (anonymous ftp, AT&T mail, commercial networks, cooperative networks, cost, electronic sources of information, network etiquette, online publishing, playing games online, services and applications, shopping services, and software copyrights), and three links out (baud, Internet, and listserv). Because there is an introductory article that contains links to at least the seven main subtopics, the top-level organization of the hyperdocument is hierarchical. Some articles are definitions or short essays that have no links out (i.e., they are leaves), except backlinks which are always present. Perusal of the out links for "0's" can quickly show isolates that cannot be reached through direct links. Note that the introductory article is one such legitimate node, and the alphabetical listing of articles is always accessible but not via an embedded textual link.

Analysis of the in links and out links demonstrated the "balance" of this corpus. There were 592 in links and 631 out links in the Netguide. In links ranged in number from 0 to 26 and there were a mean of 3.5 in links (standard deviation=3.9) per node. Out links ranged in number from 0 to 46 and there were a mean of 3.8 out links (standard deviation = 6.1) per node. A special node was created that listed definitions and this node has the most out links, with 46. The main subconcept nodes, as would be expected had large numbers of out links:

basic concepts—32
services and applications—21
specific networks—11
network issues—11
standards—11
procedures—18
resources—7

The most "popular" nodes, that is, those that had many in links were:

Internet—26
Unix—20
eMail[4]—16
file transfer—15
host—15
WAM[5] account—14
modem—13
ftp—12
bbs—12
protocol—11
server—10

These results are similar to previous designs based upon this same method of developing a vocabulary list to guide hypertext design. A former project has 68 nodes, 322 links, and 4.7 nodes per link. The network hyperdocument was more substantial both in the number of nodes and in the density of nodes (mean = 7.3 links per node),

---

4. Another node, called "electronic mail" had 14 in links. This was a design error since there, of course, should have been synonyms.

5. WAM is the acronym for Workstations at Maryland. All students at the university can obtain accounts for accessing computing resources, including Internet.

but had the same overall structure and usability "feel." The general organization is hierarchical near the top and then highly web-like after a few links to more detailed information. A 168-by-168 link plot of the database reveals general regularities, with line-like traces for main subconcept nodes and "popular" nodes. In the three years this class project has been done, the size and richness of the hypertexts has increased, due in large part to greater emphasis upon the development of a controlled vocabulary in a project's earliest stages.

CONCLUSION

The example discussed here illustrates how creating a conceptual scheme in advance can assist hypertext design. This is particularly important for group authoring projects. Given a topic, a top-level algorithm for hypertext authoring that begins with an index first is as follows:

1. Identify main facets of the topic
2. Generate an exhaustive list of terms/phrases
3. Map terms/phrases to facets, revise facets if necessary
4. Determine preferred term/concepts (label nodes)
5. Write articles (create nodes), mark cross references (links) to other nodes during writing
6. Review articles (nodes), revise node set according to criteria—grammar, style, readability, etc., links to other nodes
7. Import files into hypertext system and implement links
8. Test and edit final hyperdocument.

Critical steps in this algorithm that are specific to indexing include facet identification, listing of terms/phrases, mapping of terms to facets, identifying occurrences of concepts in nodes, and mapping those occurrences to the controlled vocabulary (i.e., linking). Note that this procedure theoretically provides an unordered set of nodes rather than a sequential document or outline. In fact, the hierarchical organization at the top of the Netguide hyperdocument is to aid authors in organizing and mapping terminology and especially to help users as they make entry into the hyperdocument and navigate through it. The ordering of traversal is arbitrary, however, guided by the user's selections during use. Therein lies the paradox of structure and freedom, the conceptual scheme and controlled vocabulary provide a structure that supports authoring, linking, and navigation while the order of traversal remains under user control. User control via linking is the essence of hypertext and the organizational interplay between nodes and links can be managed by applying principles and techniques of indexing from the start.

There is much to learn about authoring hypertexts and evaluating their design and use. Continued experience with implementations and progress in developing tools for linking and analyzing links will foster this learning. Information scientists have much to offer as the many communities involved in hypertext technology continue to make progress toward these goals.

ACKNOWLEDGMENT

The author would like to acknowledge the work of Richard Furuta in producing the link plots for the Netguide.

NOTES

Akscyn, R., McCracken, D. & Yoder, E. (1988). "KMS: A Distributed Hypertext System for Managing Knowledge in Organizations." *Communications of the ACM*, 31(7), 820-835.

Akscyn, R. (1991). *The Association for Computing Machinery Hypertext Compendium*. New York: ACM Press.

Bernstein, M., Bolter, J. D., Joyce, M. & Mylonas, E. (1991). "Architectures for Volatile Hypertext." In *Proceedings of the Third ACM Conference on Hypertext*, (pp. 243-260.). San Antonio: ACM Press.

Bolter, J. D. (1991). *Writing Space: The Computer, Hypertext, and the History of Writing*. Hillsdale, NJ: Lawrence Erlbaum.

Botafogo, R. A. & Shneiderman, B. (1991). "Identifying Aggregates in Hypertext Structures." In *Proceedings of the Third ACM Conference on Hypertext*, (pp. 63-74). San Antonio: ACM Press.

Brethauer, D., Plaisant, C., Potter, R. & Shneiderman, B. (1989). "Evaluating Three Museum Installations of a Hypertext System." *Journal of the American Society for Information Science*, 40(3), 172-182.

Burger, A. M., Meyer, B. D., Jung, C. P. & Long, K. B. (1991). "The Virtual Notebook System." In *Proceedings of the Third ACM Hypertext Conference*, (pp. 395-401). San Antonio: ACM Press.

Conklin, J. (1987). "Hypertext: An Introduction and Survey." *IEEE Computer*, 20(9), 17-41.

Crane, G. (1988). "Redefining the Book: Some Preliminary Problems." *Academic Computing*, 2(5), 6-11, 36-41.

DeRose, S. J. (1989). "Expanding the Notion of Links." In *Proceedings of Hypertext 89*, (pp. 249-257). Pittsburgh: ACM Press.

Egan, D. E., Remde, J. R., Gomez, L. M., Landauer, T. K., Eberhardt, J. & Lochbaum, C. C. (1989). "Formative Design Evaluation of Superbook." *ACM Transactions on Office Information Systems*, 7(1), 30-41.

Frisse, M. (1988). "Searching for Information in a Hypertext Medical Handbook." *Communications of the ACM*, 31, 880-886.

Frisse, M. F. & Cousins, S. B. (1992). "Models for Hypertext." *Journal of the American Society for Information Science*, 43(2), 183-191.

Girill, T. R. (1991). "Information Chunking as an Interface Design Issue for Full-Text Databases." In M. Dillon (ed.), *Interfaces for Information Retrieval and Online Systems: The State of the Art* (pp. 149-158). New York: Greenwood Press.

Halasz, F. G. (1988). "Reflections on Notecards: Seven Issues for the Next Generation of Hypermedia Systems." *Communications of the ACM*, 31, 836-852.

Horton, W. (1991). "Assay for Designers." *Technical Communications*, First Quarter, 28-33.

Jonassen, D. H. & Mandl, H. (Ed.). (1990). *Designing Hypermedia for Learning.* Berlin: Springer-Verlag.

Joyce, M. (1990). "Afternoon, a Story." In Cambridge, MA: Eastgate Systems.

Landow, G. P. (1989). "Hypertext in Literary Education, Criticism, and Scholarship." *Computers and the Humanities,* 23, 173-198.

Marchionini, G. & Crane, G. (in press). "Evaluating Hypermedia and Learning: Methods and Results from the Perseus Project." *ACM Transactions on Information Systems.*

Marchionini, G., Liebscher, P. & Lin, X. (1991). "Authoring Hyperdocuments: Designing for Interaction." In M. Dillon (ed.), *Interfaces for Information Retrieval and Online Systems: The State of the Art* (pp. 119-131). New York: Greenwood Press.

Mylonas, E. & Heath, S. (1990). "Hypertext from the Data Point of View. Paths and Links in the Perseus Project." In A. Rizk, N. Streitz & J. Andre (eds.), *Hypertexts: Concepts, Systems, and Applications.* Paris: Cambridge University Press.

Neuman, D. (1991). "Evaluating Evolution: Naturalistic Inquiry and the Perseus Project." *Computing and the Humanities,* 25, 239-246.

Nielsen, J. (1989). "The Matters That Really Matter for Hypertext Usability." In *Proceedings of Hypertext 89,* (pp. 239-248). Pittsburgh: ACM Press.

Salton, G. (1989). *Automatic Text Processing: The Transformation, Analysis, and Retrieval of Information by Computer.* Reading, MA: Addison-Wesley.

Smith, J. B., Weiss, S. F., & Ferguson, G. J. (1987). "A Hypertext Writing Environment and Its Cognitive Basis." In *Proceedings of the Hypertext 87 Workshop,* (pp. 195-214). Chapel Hill, NC: ACM Press.

Vaccaro, B. & Valauskas, E. J. (eds.). (1989). *Macintoshed Libraries 2.0.* Cupertino, CA: Apple Library Users Group.

Walker, J. H. (1988). "Supporting Document Development with Concordia." *IEEE Computer,* 21(1), 48-59.

Yankelovich, N., Meyrowitz, N., Haan, B. & Drucker, S. (1988). "Intermedia: The Concept and the Construction of a Seamless Information Environment." *IEEE Computer,* 21(1), 81-96.

Chapter 5

ONLINE HELP SYSTEMS: A MULTIMEDIA INDEXING OPPORTUNITY

Nancy Mulvany

INTRODUCTION

The past year has seen delivery begin on the promise of multimedia. The Seventh International Conference & Exposition on Multimedia and CD-ROM in San Francisco drew hundreds of exhibitors and thousands of attendees. Massive amounts of paper documents have been transferred to CD-ROM and turned into interactive multimedia documents. While I find such large-scale projects fascinating, I must admit that my professional focus is much more limited. I deal with megabytes, not terabytes, as I design index structures for information access in reference material for software and hardware products.

It is in the context of online help systems that I wish to discuss index structures and multimedia. While industry attention has been directed toward glamorous multimedia projects, it has neglected the design of efficient and useful online help systems that are accessed by hundreds of thousands of users a day. I believe that online help systems (particularly those running in the Microsoft Windows environment) are woefully inadequate, unmindful of the multimedia opportunities that currently exist, and, therefore, poorly positioned to address the future capabilities that the multimedia platform will provide. There are two primary reasons for the poor performance of current online help systems: a lack of index structures, and awkward software tools.

For many years programs have offered users access to online help. These systems were quite diverse—online help was accessed in different ways, and the user interface for online help varied greatly from program to program. The widespread use of the Microsoft Windows operating environment, however, has encouraged the design of more consistent help systems. Programs that run under Windows frequently offer a consistent form of access to the online help. Also, once the user gets into the online help, the interface is usually the same from program to program. Such standardization often contributes to greater usability of software.

Not only are most major applications shipped with online help, the online help is taking on a life of its own. There is a disturbing, but understandable, trend to refer users from printed documentation to the online help. This is disturbing because it means that the printed documentation is incomplete. A recent poll of *PC Computing* readers clearly indicated that they prefer to use printed documentation rather than online help.[1] The trend is understandable because it is far more cost effective to keep online text files current than to print new manuals. Online help was once a frill, a feature that distinguished one program from another. Today, online help is becoming the most current source of information about an application. Efficient access to this information is crucial.

INDEX STRUCTURE

Printed manuals will not become a relic despite the cost savings that could be realized! Let's look at information access in a typical reference manual for a program. The back-of-the-book index generally used in printed manuals is familiar to all of us. Computer documentation, unfortunately, offers many examples of bad indexes, although there are some good indexes as well. The problem with a good index is that it is low-profile—users jump to and from the index and the document very quickly. The structure of a good index is transparent to most users. Rarely do we stop to think, "Wow! How did it happen that I was able to find this information so quickly!"

Unless you are an indexer, it is not common to think about sound index structure. However, suffice it to say that a good index, i.e., one that works for you, is a highly structured document in and of itself. It is a document separate from the text indexed. In many ways the index is a hypertext, allowing readers to access information in a nonlinear manner. Through appropriate cross-references internal navigation aids are provided that lead to efficient use of the index and, therefore, more efficient access to information in the text. It is the index in reference books that allows us to jump into and move around in the document in a nonlinear fashion.

The *PC Computing* poll referred to above asked its readers to rate the importance of various components of documentation. For both hardware and software documentation, readers rated the index by a great margin as the most important part of the package. These are the same readers who preferred printed manuals to online help systems. Could it be that they expect *at least* the same quality of access to information in online help as they get in a book? I suspect so. I submit that until we provide the same quality of information access as that provided in a good book index, we will not be able to tap the additional power of the online environment.

The identifiable procedures that contribute to the design of a sound and useful back-of-the-book index can be applied to the design of index structures for online help systems. Furthermore, an index structure is familiar to most users—they have seen and used indexes, and the nodal hierarchies (i.e., main heading, subentries, cross-references) are understood immediately. As part of a user interface of online help, the use of an indented-style index does not require a steep learning curve.

Users access online help for only one reason. They are in a very different state of mind than, say, the user of an index in a history book or a cookbook. The online users have a problem, they want information that will solve the problem. This is neither a time for playful browsing nor for wistful exploring. Very likely, the situation represents a near crisis. Often they are aggravated to some degree because the problem exists in the first place. All they want is the information needed to solve the problem so that they can continue with their work. Efficient access to information is crucial.

In Microsoft's Word for Windows word processing program,[2] users access the online help by selecting the Help option from the main menu bar. If *Help Index* is then selected, the following options are presented:

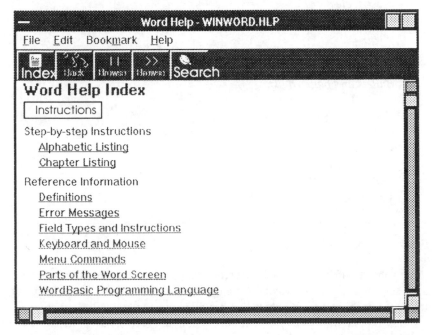

Figure 1. Word Help Index

The "Instructions" box and the underlined phrases are all linked to other screens. Of the ten options available, only one ("Alphabetic Listing") resembles the subject index usually found in a book. The "Chapter Listing" option is better known as a table of contents. Remember that the user selected *Help Index*; instead of offering an index, the system presents an intermediate screen creating an extra step in the search for information.

Next the user selects an option, usually by clicking on it with the mouse. The "Alphabetic Listing" selection brings us to another screen that lists subject entries in alphabetic order (see Figure 2). This screen can be browsed by using the vertical scroll bar on the right side of the screen; however, since this is a tedious way to move through the index, the recommendation is that users choose the Search button.

-A-
Address (labels)
Aligning text
Alphabetizing lists and text
Annotations
Automatic saving

-B-
Backup up copies
Block (selection)
Body text
Bold
Bookmarks
Borders
Bullets and bulleted lists

-C-
Calculations
Capital letters
Case (of letters)
Cells (in tables)
Centering text

Figure 2. Alphabetic Listing

All subject entries are linked to screens that usually contain yet another set of related subject entries. The second set most closely resembles the subentries in an index. It is this second set of entries that provides links to textual information.

If we formatted the *A* entries and their secondary screens as an index, it would look like the sample in Figure 3.

Let's suppose that I want information about how to right align page numbers in a footer that also contains left aligned text. I notice that there is an entry for "Aligning text" on the first screen. If I select that phrase, I am presented with a second screen that contains the entries listed in Figure 3 under "Aligning text," I will refer to these entries as subentries. At this point I am in the third screen. It has taken three steps to get here and nothing so far indicates that I will find information about right alignment of text in a footer.

If I select any of the subentries, I am taken to another, a fourth, screen. This screen contains textual information. Within these screens many terms are linked to yet other screens. Also, it is at this fourth level where we finally are shown some cross-references!

Addresses (labels)
    overview of form letters, mailing labels, and other merged documents
    attaching and filling in information in a data file
    opening and running the mailing label macro
    printing envelopes
    printing mailing labels
Aligning text
    centering text
    changing paragraph alignment
    changing vertical alignment
    formatting a paragraph
Alphabetizing lists and text
    overview
    sorting a list alphabetically or numerically
    sorting rows in a Word table
    sorting rows of text in columns
    sorting a column
    reordering paragraphs in a document
Annotations
    Overview of footnotes, annotations and revision marks
    creating, inserting, and displaying an annotation
    deleting an annotation
    Going to an annotation
    locking or unlocking a document
    pasting an annotation into document text
    printing annotations
    reading annotations
Automatic saving
    automatically saving the current document

Figure 3. Combination of Help Index Screens

No wonder that the *PC Computing* readers prefer the printed manual (and index) to this heavily-layered example of online help. One of the virtues of a printed index is that the entries are easy to scan in their entirety; cross-references are readily apparent.

The developers of this online help system could have presented an index, with main headings, subentries, and cross-references, on the first screen that appears after the user choose Help Index. Users could scan (browse) the entire index. They would likely benefit from the serendipity of discovery that users of printed indexes take for granted.

The design of this online index was clearly geared toward searching, not browsing. Once the user is in the index, a Search button is always available. Very likely users of online help systems begin their quest for information by choosing the Search option.

Had I ignored the index entries and searched for "right alignment," I would have quickly gotten results. However, none of the information retrieved dealt directly with my problem of right alignment in footers. The Search operation in this online help system is typical of many other systems. It is a simple string search, there are no Boolean operators or any other ways to narrow a search. I could not specify, for example, search for "right alignment" and "footers." Furthermore, rather than full-text availability, the search is restricted to a controlled vocabulary list that is not readily accessible to users.

Putting aside the problem of multilayered access, if we were to list the access points for "Aligning text" entry, the poor design of the index structure itself becomes apparent.

Aligning text
    centering text
        See also Paragraph formatting
    changing paragraph alignment
        See also Paragraph formatting
    changing vertical alignment
        See also Sections: Formatting Parts of a Document
    formatting a paragraph
        See Paragraph command (Format menu)
        See Storing text or graphics as a glossary entry
        See also Paragraph formatting

Figure 4. Exploded View of the *Aligning text* Entry in the Online Help

Three out of the four subentries refer the user to the "Paragraph formatting" entry. Users moving to the "Paragraph formatting" screen discover no less than sixteen subtopics (subentries) on the screen! One of the tasks of an indexer is to gather related information together. An obvious sign of an ill-conceived index is scattered information that has not been gathered together.

When I first started analyzing this particular help system, I wondered if the index entry screens contained the same entries as the manual for the product. To my surprise, the two index systems were quite different. For example, in the Word for Windows *User's Guide* we find the entry as in Figure 5.

This entry for "Alignment" is far superior to the entry in Figure 4. The language of the entry is concise and related information has been gathered together. If we compare the main headings in the printed index with the main headings in the online in-

```
Alignment
     aligning vs. framing text
     bordered objects
     buttons
          on ribbon
          on ruler
     header/footer
     paragraph
     positioned object
     rows in table
     tab stop
```

Figure 5. *Alignment* Entry from the Printed Manual

dex, the pattern of poor structure of the online index is repeated. The printed index in the manual is far more extensive than the online index. The printed index is far easier to use. Like the readers of *PC Computing*, I prefer the printed documentation and its index to that which is available online.

It should be noted, however, that the online help may not contain the depth of information found in the printed manual. One problem with the inability to locate information is that users do not know whether the information is truly not present, or whether they are not searching for it correctly.

CHARTING A BETTER COURSE

No fundamental reason exists for online help systems to be less useful and easy to use than printed indexes. However, the basic user interface for online help needs work. When users choose Help Index, that is what they should get—an index, not a table of contents.

One of the most useful features of printed indexes is the ability of users to browse through them easily. Most of us have scanned a printed index in search of a particular term only to have another entry *that we were not searching for* catch our eye. Such serendipity should be commonplace in the online environment as well. The index should be presented in its full, structured format. That is, main headings and their subentries should be visible. Cross-references should be readily apparent and linked for immediate access to the referenced entries.

The vertical scroll bar aids in speedy access to far-flung parts of the index, but moving through an index with the scroll bar can be imprecise. Part of the index screen might be composed of lettered buttons that can be used to jump immediately to a letter group, e.g., the *Ts*, *Gs*, etc. (Figure 6).

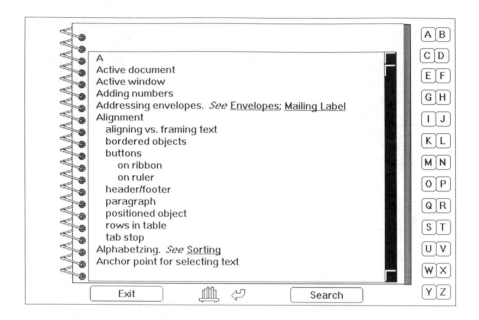

Figure 6. Sample Online Index Screen

The Search function should include the option of searching a controlled vocabulary and/or full-text. Users should have access to the controlled vocabulary. Additionally, we must move beyond simple string searches and include Boolean operators as well as proximity operators. Ideally, complex searches can be performed with the results or links to the results optionally deposited in a retrieval file.

Even if the user interface and functionality of online help systems are improved, all will be for naught if the underlying index structure is not also improved. Putting the user interface issues aside and focusing primarily on the actual index structure in many online help systems, we find them extremely lacking. Surprisingly, the deficiency is evident even with those software publishers who provide useful indexes in their printed documentation.

While many factors contribute to the sorry state of online indexes, one assumption that immediately precludes the design of sound, online index structures appears to dominate. The assumption is that moving documentation online changes the playing field; thus, online help designers are relieved of the responsibility of providing a thorough index because users can now conduct online searches.

Quite frankly, nothing has changed except the nature of the reference locator in the online index. Instead of using a page number as a locator, the indexer must now use some sort of file pointer. The information access needs of online help users and printed documentation users are the same. Only when we can provide such equivalency can we begin to exploit the additional functionality available in the online environment.

When the familiar conceptual index is used as a vehicle for information access, whether online or in printed media, the index structure must meet the same goals. One of the more cogent discussions of these goals can be found in the British Standard on indexes.[3] In BS 3700:1988 there is a section titled, "The Function of an Index." It reads (bold type added by author):

Identify and **locate** relevant information within the material being indexed.

Discriminate between information on a subject and passing mention of a subject.

Exclude passing mention of subjects that offers nothing significant to the potential user.

Analyse concepts treated in the document so as to produce a series of headings based on its terminology.

Indicate relationships between concepts.

Group together information on subjects that is scattered by the arrangement of the document.

Synthesize headings and subheadings into entries.

Direct the user seeking information under terms not chosen for the index headings to the headings that have been chosen, by means of cross-references.

Arrange entries into a systematic and helpful order.

Making search operations available, even sophisticated search algorithms, will satisfy some users. However, common sense tells us that most users are not well trained in the design of search strategies. The beauty of a useful index is that someone else has gathered together related information for the users and presented it in a helpful order. So many online help systems blatantly transfer the burden of designing information access structures to the user. As printed documentation and indexes have improved, users have come to expect swift access to information. It is no wonder that they prefer printed documentation to online documentation as seen in the *PC Computing* survey.

AND WHAT OF MULTIMEDIA?

Today we see mass market PC systems shipping with CD-ROM drives. The price of adding a CD-ROM player to existing systems has fallen and will continue to fall. Microsoft is including many multimedia extensions in Windows 3.1. Although developers cannot yet assume that every PC has a CD-ROM drive, as prices fall and the industry standardizes on hardware specifications, a CD-ROM drive may very well become a standard component.

In the realm of online help systems, the possibilities offered by multimedia are exciting indeed. For example, the excellent tutorial provided on a video tape by Corel

Systems for its popular drawing program, Corel Draw!, could be digitized and viewed by users online. This type of access to the tutorial would be more immediate than watching the video tape on a conventional television.

Online help designers need not wait for a CD-ROM player in every PC to try their hands at providing direct access to multimedia information. The tutorials included with many programs need their attention now. The growing dominance of the mouse-driven, graphical user interface has created a situation where instruction is often best shown, not written. For example, a spreadsheet user who wants to know how to apply a formula to a column would very likely benefit from a self-running demonstration of the procedure rather than a textual description. Such information is already present in many applications—it is found in the online tutorial.

Unfortunately, most online tutorials are not thoroughly integrated with the online help system. Usually, they are accessed separately and presented in a linear fashion. It is often difficult, if not impossible, for users to jump in and out of sequences in the tutorial. However, if the material were indexed, users could access discrete portions of the tutorial.

The online index could offer users a choice between a textual description of a topic and a demonstration of how to accomplish a task. If we add to this scenario access to discrete segments of digitized video, the opportunity for presentation of helpful information is greatly enhanced.

If, however, developers fail to provide rich, structured indexes in the multimedia environment, users will continue to be disappointed. If it happens that current, reliable information can only be found online, not in printed documentation, efficient access to online information will become critical.

SOFTWARE TOOLS

To construct efficient and thorough online help systems, developers must often use several programs rather than one integrated design system. While much of the printed documentation is indexed using embedded indexing software tools, the documentation text files and their indexes are too often not seamlessly portable to the online environment. The problems with embedded indexing software tools that are used to produce indexes for printed documentation have been documented elsewhere.[4] Indexers wishing to index online material are thwarted to even a greater degree than writers who embed index entries in text files for printed documentation.

Ideally, text that originates in the files used for printed documentation can be moved online easily, along with pointers for the index entries. Since the pointers in the files for index entries have already been created for the printed documentation, the preexisting index structure should be useful for the online help system.

Additional material will often be included in the online index, such as references to tutorials and video sequences. The online indexer will need tools that can be used to integrate the existing index for the text files and new entries for online material.

The design of efficient and useful index structures for any medium, whether online or print, is a complex and demanding task. Developers of online indexes will find it difficult to achieve the goals outlined above in the British Standard as long as the software tools obstruct sound index construction. The lack of sophistication found in so many embedded indexing software tools for print media does not bode well for the indexer working solely in the online environment. As more critical information goes online, users' access needs are of greater concern. Opportunities for providing access to information in multimedia environments await us—but we will fall far short of exploiting the potential of this rich environment if we lack the development tools needed.

NOTES

1. Grech, Christine. "Computer Documentation Doesn't Pass Muster." *PC Computing*. April 1992, 221-214.
2. *Microsoft Word for Windows 2.0* (1991). Redmond, WN: Microsoft Corporation.
3. *British Standard Recommendation for Preparing Indexes to Books, Periodicals and Other Documents* (BS 3700:1988). London: British Standards Institution, 1988.
4. Mulvany, Nancy. "Software Tools for Indexing: What We Need." *The Indexer*, 17(2), 108-113.

Chapter 6

HYPERTEXT AND INDEXING

Peter Liebscher

INTRODUCTION

Hypertext has fired the imagination of both scholars and information professionals. Vannevar Bush's Memex, a comprehensive, personal database intended to supplement a scientist's memory, was an early vision without a technology.[1] Today's technology promises to revolutionize the way we store, find and exploit information.

Easy access to Hypercard for Macintosh users and Windows-based hypertext systems such as Guide have encouraged a host of information providers to enter the market with hypertext products. Unfortunately, many of these products fail as effective search and retrieval systems because their producers have not grasped the importance of sound conceptual organization for hypertext. There are fundamental problems of organizing a hypertext for access and retrieval. How should a hypertext be structured? What access methods should be provided and how should these, in turn, be structured? Who should organize a hypertext—authors or "hypertext" professionals?

In many ways, the task of organizing hypertext is strikingly similar to traditional indexing. It can be argued that hypertext links and index terms have much in common. Links provide meaning and structure to a hypertext. Indexes can do the same for other documents. This being so, the corpus of knowledge developed and applied effectively by indexers over the years can be applied to organizing hypertext. Perhaps indexers should construct hypertext.

This chapter's intent is not to put forward a solution to the problems of organizing hypertext for retrieval. Rather, it aims at identifying the many similarities that exist between creating indexes and creating hypertext and suggests that hypertext producers ought to look at the work of indexers for solutions to some of their problems.

HYPERTEXT

The simplest definition of hypertext is that of a set of concepts connected by links, where the links are relations. Smith and Weiss point out that hypertext, like

Bush's Memex, is distinguished from other forms of data representation by an associative structure that closely models the structure of human memory.[2] Whether that analogy holds true or not, hypertexts do consist of nodes connected by links according to a set of organizing principles that make up the associative structure. A browsing tool is usually provided so that users may traverse the links. Within a hypertext, the presence of a link may be indicated through symbols such as a word, a phrase, an icon, or other representation. These symbols can be said to represent the "meaning" of the new node being linked. For example, the highlighted term *air pollution*, embedded in a node on pollution, indicates a link to a node on air pollution and thus carries the "meaning" of that node.

But, with the exception of a browsing tool, all documents (books, journal articles, technical reports, etc.) meet this criterion. Print documents contain links that are both explicitly flagged (e.g., *see page 24*; *see Table 1*; *(Marchionini, 1989)*) and shown implicitly (e.g., a section is conceptually related to the sections that precede it and the sections that follow). In a print document, most of the links are indicated implicitly in its sequential structure. Far fewer links are explicit. In hypertext, on the other hand, explicitly flagged links are far more prevalent and may even outnumber implicit links.

INDEXES

Good indexes provide efficient access to information carrying materials such as text, graphics, sound, etc.. Heilprin describes the relationship of an index term to the material being indexed as a form of homomorphic reduction or, in his words, a "paramorphism"; i.e., mapping from a complex pattern such as represented by a book to a simpler one such as an index term.[3] Thus we have a relation between index term and document. A single index term may map to a section of a document, to a complete document, or to an entire collection of documents. If the index term maps to a document, it represents the "meaning" of that document. A set of index terms—an index—represents a document or set of documents, in terms not only of meaning but also of structure. The index has mapped onto it one of many possible semantic structures of the document.

An index may be viewed as a document's much abbreviated description or even as its surrogate. As such, a table of contents can be described as an index. While tables of contents are normally thought of only as entry points to documents, they also have mapped onto them the content and structure of their associated documents and are, therefore, document surrogates. If the structure of a document is mapped onto the table of contents, then a virtual table of contents must be embedded in the document. This holds true not only for the special case of a table of contents, but for any topically organized index to a document. Consequently, documents contain, embedded within themselves, one or more virtual indexes.

One normally thinks of an index as following the document in time. First the document is authored, then indexes are produced, not necessarily by the author of the document. Let us assume, however, that authors develop documents from outlines, even if these exist only in an author's mind. We can then think of these document outlines as virtual indexes. Where such an "index" does not have a physical form, it can, nevertheless, be viewed as a nascent index that can take a physical form if the author desires. As soon as the document has structure, even if only in the mind of the author, the document has an index. Assuming that a document is not written strictly sequentially, the author may use this index to further refine the document's structure and as a finding tool for specific concepts in the document. The index, like the document it represents, is a living, dynamic object that grows as the document grows. The relationship is symbiotic. Changes in the index change the document and vice versa.

Of course, other indexes may be created after the document has been completed as access points to concepts that were not part of the author's original conceptual structure. Many "back of the book" indexes are created in this way. A document may have several physical indexes, some created with the document, some after its completion. But all these indexes are implicit in the structure and content of the document. To claim that documents contain their own indexes is, therefore, certainly reasonable.

The concept of virtual, embedded indexes is important when looking at the similarities between indexed documents and hypertext. As long as we view an index only as an entity separate from its document, we obscure the similarities. If, however, we think of an index as an organizing structure for a document, the similarities become more apparent.

INDEX TERMS

Can we argue that index terms and hypertext links are the same animal? A link is a true relation and, together with its associated concepts, can be expressed in terms of a predicate calculus. While links express relationships between two concepts, an index term is a concept that is linked, by a separate mechanism, to another concept. A link in an index is not part of the index term. For example:

Ozone layer 324

is made up of an index term, *Ozone layer*, and a link. It would be wrong, however, to think of either *324* or *Ozone layer* as a link. They merely flag the presence of a link. The real link is the implicit relation *represents*. So what we actually have is the paramorphism:

abbreviated concept *ozone layer* **represents** expanded concept *ozone layer*

where the number *324* is merely a physical pointer to the location of the expanded concept. What we consider to be a link should really be viewed as a mapping between two concepts and that this mapping is physically enabled through a device such as a page number. There is no explicit representation of the relation in the index.

There are links in an index other than those that map documents to index terms. These links express the relationships between index terms. Such links can be explicit, e.g., "*see also* pollution." They can also indicate a number of different relationships, both hierarchical and associative, e.g.:

```
air pollution    BT  pollution
                 NT  smog
                 RT  water pollution
```

However, more often they are implicit. For example, in a topically organized index, the relationship between index terms is often indicated through the relative position of the terms in the index. This relationship can be quite strong. So strong that it can be argued that transforming an index term in an hierarchical index to a random selection of characters will not always rob it entirely of its meaning.

```
Aircraft
    Lighter than air
        Balloons
        Dirigibles
    wb%*j12>
        piston engined
        turbojet
        rocket
Ships
```

This example illustrates that a considerable amount of residual meaning is retained by the random characters, wb%*j12>, that replaced the term originally in that position.

HYPERTEXT LINKS

Shepherd & Watters describe a hypertext link as:
A labeled pointer from one node to another indicating some type of relationship between objects associated with the nodes. Explicit links may be "next", "notes-on", "author", "earlier version", and so on. Implicit links may also be present such as links to dictionaries which are always there.[4]

While their definition is certainly mainstream and one would not expect much argument, it does obscure the real nature of a hypertext link. It is the presence of a link that is indicated in a hypertext by words, phrases, icons, or "hot spots" on the screen. Nonetheless, these symbols are not themselves the links. As with traditional printed indexes, hypertext relationships are only implicit in what are commonly called hypertext "links." Thus, while clicking on the embedded term "ozone layer" in a hypertext may bring to the screen information on the ozone layer, the true relationship here is just as in the printed index:

abbreviated concept *ozone layer* **represents** expanded concept *ozone layer*

An electronic pointer is the physical device that makes the connection but the real link is the relation **represents**.

Just as links in a traditional index can specify the type of relationship that exists between index terms, hypertext links can indicate relationship types between a "link" symbol and the text that is being linked. Such links are known as *typed links*. Typed links are used to provide more information about the destination of the link.[5] Thus, typed hypertext links can distinguish between, for example, hierarchical relationships and purely associative ones or even the type of hierarchical or associative relationship.

WHAT ARE THE DIFFERENCES?

How then, do index terms and hypertext links differ? We have argued that an index is embedded in all documents, including hypertext. We have also argued that "indexes" in both hypertext and other documents contain links, but that these are hidden relations that map an index term, an abbreviated concept, to another, expanded concept. Both hypertext authors and indexers determine relationships between concepts in their respective documents and then select appropriate symbols, e.g., index terms, icons, etc., to indicate the presence of links. What are commonly, but erroneously, termed links in a hypertext, are really embedded index terms that derive their meaning partly through their paramorphic relationship to expanded concepts in the hypertext and partly through the relationship derived from their position in the hypertext.

Further, we have argued that a printed document contains its own index. It may be separate and explicit, as in a table of contents. It may be embedded in the document yet explicit (chapter and section headings). Finally, it may be implicit (expressed in the structure of the document). Is this true for hypertext? Certainly a hypertext may have associated with it separate indexes as access points for retrieval (Figure 1). A hypertext may also have embedded within its text terms that define the structure of the hypertext. An example drawn from a simple hypertext illustrates the index embedded in the hypertext (Figure 2).

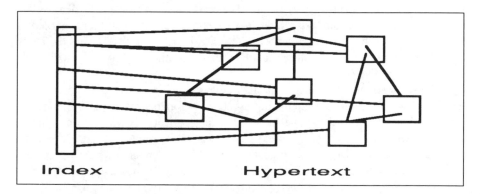

Figure 1. Graph of hypertext with external index.

Figure 2 shows that a hypertext, like other documents, contains its own "index." The terms used as "links" can certainly be copied to a separate document and ordered to reflect their position in the hypertext. As with an index to a traditional document, the relative position of the "links" in a hypertext also give them meaning. Finally, a hypertext contains a structure that is implicit in the linear arrangement of text on the screen. This too could be abstracted and made explicit through a separate index reflecting the linear structure, although usually there is little utility in that.

Pollution
Text text text text text text
text text text <u>water pollution</u>
text text. Text text text text
text text.
Text <u>air pollution</u> text text
text text text text text
text. Text text text text text

Water Pollution
Text text text text text text
text text text text text text
text text. Text text text text
text text.
Text text text text text text
text text text text text
text. Text text text text text

Air Pollution
Text text text text text. Text
text text text text.
Text text text text text text text
text text text.
Text text text <u>ozone layer</u> text
text text text. Text text text
text text text text text.

Figure 2. Embedded "links" in a hypertext.

A traditional document can have multiple indexes, each reflecting a different conceptual organization of the topic. Practically speaking, however, the document itself can have only one physical structure. The reader can, of course, physically alter the document to reflect other structures (a book can be torn apart and reassembled in a different order) but this is usually not an option with library materials. Cut-and-paste functions can be used to alter the structure of an electronic text, but this is a difficult and time consuming process.

Hypertext, on the other hand, through devices such as webs and tours, allows easy transition from one conceptual representation of a document to any number of others. While a traditional document can only change its logical structure through use of surrogates (its indexes), a hypertext has no such constraints. Nevertheless, both hypertext and traditional, indexed documents can have a number of logical structures. Of course, the user still has the problem of determining the most appropriate structure for a task, but once determined, traversal of the hypertext is easy.

CONCLUSION

We thus see a difference only in the physical structural possibilities of hypertext and traditional documents. The one can be displayed directly in a number of different conceptual structures; the other must do so through surrogates. Is the difference fundamental? It is, certainly, for the user who has available in a hypertext options that are difficult or impossible in other representations. In terms of authoring, however, the intellectual effort is the same. Because links can be forged easily in the electronic environment of hypertext, the temptation has been to employ less rigor in determining appropriate structures for users and tasks.

In all retrieval systems, the fundamental problem has always been one of determining a sound conceptual schema that reflects the accepted structure of the topic, the user's understanding of the topic, and the user's task. This is no less true of hypertext. Where a hypertext is created from an existing document or documents, the "authoring" process is, in reality, an indexing process. The author/indexer determines the overall structure(s) of the hypertext, selects appropriate "link" representations, and creates the appropriate electronic links. This is, conceptually, no different from indexing traditional documents, where the indexer determines the structure and content of the document, selects appropriate index terms, and appends a pointing device. Where a hypertext is created de novo, the author has primary responsibility for its structure(s). Even here, however, the task is one of indexing and the assistance of a professional indexer trained in that subject area may be invaluable.

NOTES

1. Bush, Vannevar. "As We May Think." *Atlantic Monthly*, July 1945, 101.

2. Smith, J.B. & Weiss, S.F. "Hypertext." *Communications of the ACM*, 31(7), 816-819.

3. Heilprin, L. B. "Paramorphism Versus Homomorphism in Information Science." In: *Toward Foundations of Information Science*, L.B. Heilprin (ed.), White Plains, NY: Knowledge Industries, 115-136.

4. Shepherd, M.A. & Watters, C. "Hypertext: User Driven Interfaces." In: *Interfaces for Information Retrieval and Online Systems,* M. Dillon (ed.), New York: Greenwood Press, 1991. 159-167.

5. Akscyn, R.M., McCracken, D.L. & Yoder, E.A. "KMS: A Distributed Hypermedia System for Managing Knowledge in Organizations." *Communications of the ACM*, 31(7), 820-835.

Chapter 7

INFORMATION STRUCTURE MANAGEMENT: A UNIFIED FRAMEWORK FOR INDEXING AND SEARCHING IN DATABASE, EXPERT, INFORMATION-RETRIEVAL, AND HYPERMEDIA SYSTEMS

Dagobert Soergel

INTRODUCTION

Database systems, expert systems, Information Storage and Retrieval (ISAR) systems, and hypermedia systems have their distinct functional capabilities and strengths. Yet on an abstract level these systems have more in common than meets the eye—functions that on the surface seem different may, in fact, be essentially the same. This commonality is demonstrated by the development of intelligent databases and by the use of hypermedia approaches in the interface to expert systems. The journal *Expert Systems With Applications* issued a call for papers for a special issue *Expert systems integration with multimedia technology*.

What if we could design a system that would draw on the commonalities, combine and integrate all the different functions, and, through that integration, enhance them to provide great power with an intuitive interface. This is exactly what this chapter begins to do. It sets out to demonstrate the essential unity of these different systems and to develop a unified framework for improved design. It defines a system structure and generalized search and inference operators that are applicable to any type of searching—by subject, by citation, by prerequisite, or by any other criterion—following any kind of link and thus making searching both simpler and more powerful.

There appears to be no name for this type of general system. Rather than using an awkward phrase such as database/expert/retrieval/hypermedia system, we will use the term *information structure management system*, emphasizing the key element of structure common to all types of systems encompassed in the unified approach proposed.

First, we need to show that distinctions often made, particularly the one between traditional (bibliographic) retrieval systems and typical hypermedia systems, are multidimensional and a matter of degree. Then we must define a general structure

111

and data model for information structure management systems encompassing database systems, expert systems, ISAR systems, and hypermedia systems. This provides the basis for discussing search and navigation and system construction or indexing. We then revisit the unity theme by analyzing the multiple functions of objects (entities, nodes) and relationships (links) in an information structure management system. Finally, this chapter considers some issues of system design, especially building a user interface that would provide the power of the general approach without overwhelming the user.

One useful strategy in developing and presenting the unified approach is to find examples from one context and then generalize the principle to other contexts. For example, consider Boolean searching, a common function in ISAR systems. In hypermedia systems, it might be quite useful to extend navigation, which leads from a starting document to a citing document or a supporting document, to a Boolean search, where the user could start from two documents and look for documents that cite both (known as co-citation) or that support both. The examples have been chosen with a view to demonstrating the richness and power of the proposed structure.

Some of the ideas elaborated in the following sections have been mentioned or alluded to in Halasz' [1988] important paper, in which he discusses seven issues for the next generation of hypermedia systems and mentions the "unity of systems" [my term] theme. Croft [1989] and Thompson [1989] also illustrate the integration of hypermedia and retrieval. Frisse and Cousins [1992] review hypertext models with a view to integrating semantics and information retrieval features.

THE MULTIDIMENSIONAL DESIGN SPACE FOR INFORMATION STRUCTURE MANAGEMENT SYSTEMS

This section shows that database and expert systems, retrieval systems, and hypermedia systems reside not in different worlds but in overlapping regions of the same world. Concentrating on the difference between a prototypical (bibliographic) retrieval system and a prototypical hypermedia system, it analyzes the multiple facets or dimensions that characterize systems or, more precisely, individual searches, and shows that these dimensions can vary independently from each other, spanning a multidimensional design space. [Many of the ideas in this section were stimulated by conversations with Peter Evans, a doctoral student at the University of Maryland.]

A few introductory examples are useful as a backdrop for the general discussion. In preparing a lecture on a new topic, I might first do a literature search to obtain a list of documents and pick one to read first, or I might remember a document that would give a good start. The starting document might lead to a document cited, or it might indicate that I must learn about another topic first and do another search; if no such leads exist, I simply proceed to another document from the retrieval list.

The readers of a book might follow the king's advice: "Begin at the beginning and go on till you come to the end: then stop," reading one chapter after the other, one paragraph after the other. Some readers, however, might deviate from the author's sequence, consulting, for example, the table of contents or the diagram showing the logical dependencies between chapters found in the front of some books. Others might deviate from the linear path by looking at some of the footnotes or by following a cross-reference.

Although a few people read the newspaper from beginning to end, most of us pick certain sections, scan the headlines, read the first paragraph of a story and perhaps go on in that story, or jump to the next headline. In an electronic newspaper the reader can enter an interest profile to initiate a search that returns a list of stories or articles in order of importance or grouped by subject.

In each of these examples, the search for information evolves in a series of steps. In each step the user must find the next stopping point—the item of information to be consulted next. An ISAR/hypermedia system must support the user (used here in the broadest sense, including both searcher and reader) in this retrieval task.

An item of information can be any text object—from a book to a sentence—or any other media object, such as a picture, a map, or a segment of sound or a moving picture sequence that can be replayed. *Media object* is a broad term that covers all of these items; specifically, many media objects consist only of text (*text objects*) or contain a large proportion of text. The general framework presented here deals more broadly with any type of *object*, *entity*, or *item*. An object is also called a *node* when the emphasis is on the place or role of the item in a network of links. In the literature a media object is also called *frame* (in the sense of all the information fitting on one screen, not to be confused with frames used for knowledge representation), *notecard* or, simply, *card*.

Reading a whole document, such as a book or journal article, represents a special case of search. The question of what chapter (or section, or paragraph) to read next is seldom posed explicitly: by default the reader goes on to the next chapter (or section, or paragraph). But there exists a retrieval problem nevertheless, and a book solves this through its arrangement, its table of contents, or outlined as in flow chart form as mentioned above.

A book made available as a hypermedia base makes the user (reader) aware of choices and suggests detours or radical deviation from the default sequence. It does so by supporting retrieval through a network of links among paragraphs (or sections, or chapters). This kind of retrieval is an integral part of reading. (Depending on maximum node size, a hypermedia system could treat an entire journal article or even an entire book as one unit linked to other units, but that would not be in the true "hypermedia spirit.") On the other hand, a database of full-text journal articles in which the user can jump from a reference in the bibliography of Article 1 directly to the referenced Article 2 is feasible and useful, while transforming all these articles into a fine-grained hypermedia base would involve a huge effort.

On the other hand a user seeking information and faced with the question of what entire document to read next can turn to a bibliographic ISAR system. There the user cannot start from a paragraph just looked at—most systems require a query formulation in terms of subject descriptors or author names, etc.

The basic retrieval operation is always the same. Starting from a known object—a media object or a descriptor—the user follows links provided by the system to find one or more other objects. On an abstract level, using a known media object as a starting point to find descriptors for further searching is the same thing as using a known descriptor as a starting point to find media objects for perusal. Even Boolean searching works both ways: the user may start from two descriptors and look for all media objects that can be reached from **both** descriptors, or start from two media objects and look for descriptors that can be reached from **both** of these objects. (Descriptors in common to two relevant media objects are likely to be good descriptors for further search.) Moreover, as will become clear below, the same principles apply for searching in a database system that deals with any type of object, such as food products, and that uses relationships or links to store data about these objects.

Many systems today offer both links from one media object to another and links from descriptors to media objects. These are best viewed as unified systems with the same search operations usable for any type of starting object and link. Frisse and Cousins [1989] offer a different view, dividing the system into a document space and an index space.

Stressing commonalities is not to ignore or trivialize the very real differences between existing systems and how they are searched. Indeed, many ways exist for finding the next item of information and for organizing the entire pattern of search. We need a systematic analysis of these many ways of searching. This approach is summarized in Table 1, and Table 2 shows the dimensions that span a space in which any search can be located.

Table 1. Parameters or Dimensions for Analyzing Searches and Systems

"Traditional" information retrieval systems and hypermedia systems are designed to support certain kinds of searches.

The differences can be analyzed along a number of dimensions shown below.

These differences are a matter of degree.

An integrated information structure management system should support all types of searches, adapting to user needs and preferences.

From a user's point of view the best system is one that supports any type of search equally well, no matter where it falls in the multidimensional design space defined in Table 2. As a practical matter, most systems support some types of searches and some search features better than others; in that sense, the dimensions discussed can be considered as dimensions for the classification of **systems**.

Table 2. Dimensions for Analyzing Searches and Systems

This table is deliberately oversimplified. See the text for a fuller discussion of many points.

Dimension	Typical hypermedia search	Typical bibliographic search without interaction
Type of starting object(s)	A paragraph, a picture, an audiovisual object	A search key: A subject descriptor, person, organization, etc.
Method of finding good starting objects	Natural encounter during a search	Deliberate selection
Role of starting object in the entire search	Mostly objects of value in their own right	Mostly objects used only for searching
Method of specifying starting object(s) and link types	Object currently examined as default starting object Selecting link type from display on the screen	Entering elements from the keyboard
Type of target objects found	Full text, picture, etc.	References to documents
Role of the results of a single search step in the entire search	Piece of a mosaic being built Stepping stone for further searching	Final answer set of documents to be read
Number of search steps in an entire search	Many	Few
Number of objects looked for or found in one search step	Few (1 - 5)	Many (10 -20 and up)
Number of objects examined upon retrieval before moving on to the next search step in an entire search	Few	Many
Granularity: Size of objects looked for or found	Individual paragraphs, images, or sound objects	Whole documents
Completeness and complexity of search specification for each search step	Partial, often implicit Simple	Complete, often carefully worked out Complex
Ability of the user to augment the data base	Often built in as an essential part of the system	Usually no-existent

Each of the dimensions in Table 2 merits a few comments.

Type of starting object(s). A search step always proceeds from a known starting object to one or more unknown target objects. A typical hypermedia search starts from a paragraph, a picture, or an audiovisual object and follows links to other such objects. A typical bibliographic search starts from a subject or author and follows links from these to whole documents.

Method of finding good starting objects (and) **Role of the starting object in the entire search**. In the typical hypermedia search, the user looks at objects that actually provide some of the information needed and uses these same objects as starting points for the next search step. In a bibliographic ISAR system, the starting object—a subject descriptor or author name—is of interest only as a means to find target objects. A user must deliberately search her own memory or a thesaurus to find useful subject descriptors to start the search. In the prototypical hypermedia search, the user never worries about finding starting objects because the items of information read or looked at will also lead to further interesting items. In the typical bibliographic ISAR system, finding the right starting objects becomes a major concern.

Method of specifying starting object(s) and link types. The most common methods for specifying starting object(s) are 1) highlighting elements on the screen (often a default value) and 2) keying in their identifiers. From a practical standpoint, the ability to select choices by highlighting values displayed on the screen is essential for attaining the functionality of hypermedia systems, but it is neither theoretically necessary for hypermedia systems nor limited to them. Thus, this is not so much a matter of the basic nature of a system as of the mechanics of the user-system interface.

Type of target objects found. The prototypical hypermedia search leads to the media objects themselves, the full text of a document (or, more often, a paragraph or page-size unit), an actual picture, etc. The prototypical bibliographic search leads to references to documents; the user then has the added task of finding the actual text, often in paper form from a library. But some retrieval systems function much like the prototypical bibliographic system, yet give access to full text. Some systems, especially CD-ROM systems, provide a linkage between a bibliographic system and a largely unstructured repository of machine-readable documents that can be called up—a kind of electronic document delivery. On the other hand, a hypermedia system usually does contain nodes that are bibliographic references, such as an abstract of an entire document, with the document itself being stored in the hypermedia base as a group of smaller text objects.

Many databases or information systems deal with objects that cannot be stored as such in electronic form, such as food products, organizations, or persons. In that case, an electronic information system by necessity contains only references. But a storage room in which food products, technical parts, or whatever are kept in an organized fashion also represents an entire information system or a component of one.

The following two dimensions express the degree to which a search is *interactive*. For purposes of contrasting systems, this discussion treats the prototypical hypermedia search as highly interactive and the prototypical bibliographic search as one-shot, noninteractive. Of course, many bibliographic searches **are** interactive—proving the point that the distinctions between hypermedia searches and bibliographic searches are multifaceted and a matter of degree.

Role of the results of a single search step in the entire search. What does the user intend to do, or actually do, with the search results? The user might: 1) consider the results of one search step merely as a piece of a mosaic being built through many steps—the "berrypicking" approach to assembling information typical of hypermedia searches (The term "berrypicking" coined by Bates [1989] in a paper suggesting a highly interactive and incremental method for bibliographic retrieval); 2) use the object(s) found as steppingstones for further searching, without or in addition to using them as sources of information in and of themselves, also fairly typical of hypermedia systems; 3) look at the results of a search step as the final answer set containing all the objects to be considered in assembling the needed information—a situation typical of bibliographic searches without interaction.

Number of search steps in an entire search (information seeking episode). This dimension is closely related to the intended use of target objects. The typical hypermedia search consists of many search steps; in the prototypical bibliographic search there is just one step.

Number of objects looked for or found in one search step. In the typical hypermedia search, the user seeks just the next object to look at—following links from the object being examined generally leads to one or a few other objects. In the typical bibliographic search, the user is looking for a list of references, perhaps 10 or 20, to be used as the final answer set; a search may retrieve many more.

Number of objects examined upon retrieval before moving on to the next search step. The user of a hypermedia system often examines only the first of several retrieved objects and immediately uses that object as the starting point in a next search step. The user of a bibliographic search typically examines all documents selected as relevant from the retrieved set and initiates additional search steps only if the information received is not sufficient or points out another information need.

Granularity: Size of objects looked for or found. The typical hypermedia system allows very specific access to individual paragraphs, images, or sound packages (such as a definable unit within a piece of music). The typical bibliographic ISAR system supports only searches for whole documents. (Sometimes book chapters may be treated as individual documents.)

Completeness and complexity of search specification required in each search step. The definition of a search step developed so far—starting from a single starting object and selecting all target objects reachable by a given type of link—is oversimplified. While this simplicity will do for some searches, others use a much more complex search specification, such as a Boolean query formulation, a set of descriptors used in a similarity search, or a query formulation using a relationship with multiple arguments. Devising such a query formulation requires effort. The typical bibliographic ISAR system works best with a complete and carefully worked out search specification. In the typical hypermedia search the user is not even aware of using search specifications in selecting a link type for a search starting from the object being examined. Put differently, a search may consist of one very complex step or of many simple steps.

117

Ability of the user to augment the database. This dimension, also, concerns a more technical issue. The prototype hypermedia system allows the user to enter new objects and new links to be added to the public database or to the user's private version. This is in keeping with the idea that searching and the user's own effort of reading, note-taking, commenting, and writing are not separate activities, but integrated in the interaction with a hypermedia system. Allowing user feedback to update and improve the database would also be a great benefit in bibliographic and fact retrieval systems, but that feature is much less common there.

The dimensions just discussed are to some extent independent from each other. For example, the Information Navigator by IME is a bibliographic retrieval system with a hypertext "feel" (it is advertised as a hypertext application). Like any bibliographic ISAR system, the Information Navigator displays bibliographic records consisting of a number of fields on the screen. (The underlying data store is arranged according to the entity-relationship approach, but that is not important for the discussion here.) Unlike other bibliographic ISAR systems, the Information Navigator lets the user select any object (person, organization, subject, series title) displayed on the screen as the starting point for a new search. Thus, a user who found a document in a subject search and wants to find more documents by the same author, needs merely to highlight the author's name on the screen and press the search key. If one of the documents thus retrieved has an interesting subject descriptor, the user needs merely highlight it and press the thesaurus search key to find broader, narrower, and related descriptors. Merely highlighting one of these and pressing the search key starts another subject search.

The remainder of this chapter sets forth a unified system structure that can accommodate any type of object and any type of relationship or link and thus any kind of data, and generalized search operators that operate on that structure. It also introduces the concepts of **neighborhood, query,** and **path** and **script**; a neighborhood is a collection of objects and also a generalization of the concept **object** since in many contexts neighborhoods can function as objects. A query can be seen as an object that defines a neighborhood. A path is a neighborhood in which the constituent objects are arranged in a sequence, and a script is an object with instructions that orchestrate the presentation of material and the user's interaction with the system.

THE STRUCTURE OF A UNIFIED INFORMATION STRUCTURE MANAGEMENT SYSTEM

An information base (defined to include database, knowledge base, bibliographic database, and hypermedia base) is made up of **objects** (also called **entities** or **nodes**), **relationships** (also called **links**), **neighborhoods** (also called **regions, clusters,** or **virtual composites**), **queries,** and **paths** and **scripts** (Table 3). The database literature uses the terms entity and relationship, the hypermedia literature uses object and link.

A neighborhood is any group of objects together with the links that exist between them. More generally, the elements of a neighborhood can in turn be neighborhoods.

Table 3. Elements of Information Structure

    Objects (entities, nodes)

    Relationships (links)

    Neighborhoods

        A neighborhood is any group of objects and/or neighborhoods, particularly a group identified through relationships or links with one or more other objects or neighborhoods, together with the relationships or links that exist among the members of the group. A search results in a neighborhood .

    Queries

        A query defines a neighborhood.

    Paths and scripts

        A path is a special type of object or node that defines a neighborhood of other objects and specifies a sequence of these objects. A script is an object with instructions that orchestrate presentation of material from an information base.

Usually a neighborhood is a group of objects or neighborhoods that have something in common, in particular a group of objects or neighborhoods identified through links with one or more other objects or neighborhoods. A search results in a neighborhood. A query specifies a search and thus a neighborhood; that specification is dynamic—the neighborhood contains the objects meeting the search criteria at the time the query is invoked. A path is a special type of object or node that defines a neighborhood of other objects and specifies a sequence of these objects. A script is an object that contains instructions that orchestrate the display of other objects. A path could be seen as a special kind of script. Both guide the user through an information base, possibly in a very prescriptive manner.

    As the following examples illustrate, almost any kind of information can be represented through proper choice of object types and relationship types

Objects (Entities, Nodes)

    A standard hypermedia system covers media objects—paragraphs, pictures, sound objects—without differentiation of object types, and many systems do not differentiate between link types either. In the words of Halasz [1988]:

> To the system, all nodes and links are essentially the same: objects to be stored, retrieved, displayed, interconnected, etc. To the user, nodes and links are filled with meaningful contents and organized into meaningful structures. The system cannot operate directly on

this meaning. It simply provides a collection of generic tools that can be used to manipulate networks in a meaningful way.

The future, however, belongs to systems that deal with meaning and can thus provide more intelligent support. For example, Carlson [1990] presents a system that deals with strategies, objectives, and issues and their causes and allows for sophisticated retrieval and processing through its rich semantics of object types and link types. While on the surface each object in this system may look just like a text object, object types are distinguished by what the text represents. Similarly, assertions, such as assertions on the status of some situation in the world or mathematical theorems, can be treated as objects in a hypermedia base/database, where each assertion can be represented as a text object or in a formal entity-relationship representation as discussed below.

A typical database covers objects such as persons, organizations, food products, or technical products and uses relationships to express data about them. A generalized database/hypermedia base structure can cover a variety of object types (see Table 4 for examples).

Table 4. Object Types. Examples

Media object (text, graphics, sound) of any size	Person or organization
Path	Concept (subject, topic)
Database, data set	Computer program
Assertion	Organism
Problem	Food product
Strategy	Building
Objective	Work of art
Issue	Technical product, device (anything from a screw to an engine to an entire airplane)
Situation, circumstance (which may be the cause for an issue)	Person

Relationships (Links)

To be useful, an information base must make statements about the objects it covers. Such statements consist of a relationship with one or more arguments, and with each argument slot filled by a specific object value; e.g., *Smith 1991 supports Miller 1990* (a statement about two media objects), or *Brownie has-ingredient Chocolate* (a statement that makes an assertion about the world directly). A more complete form of the second statement, using a relationship with six arguments, is *Brownie has-ingredient [Chocolate, rank 3, percent of weight 20%, percent of dry 25%, (purposes: improving taste, chewy consistency)]*. An information base is a collection of such statements.

Many relationships have two arguments, they are binary. A binary relationship can be considered as a **link** between its two arguments. Hypermedia systems use binary relationships and use the specific term "link." Database and expert systems use many binary relationships but also relationships of higher order and appropriately use the more general term "relationship." In this chapter we use "link" when the emphasis is on binary relationships, particularly in the context of using these relationships for navigation in a hypermedia base. We use the more general term "relationship" whenever the context requires it.

Each link could be assigned a linking strength. In subject indexing this is known as weighting. These weights can be used in searching to compute a measure of expected relevance.

Table 5 gives examples of relationship types. These are discussed below to illustrate the wide range of information that can be handled in a unified information structure.

Some relationships can be applied to many types of objects. For example, *produced-by* can link any type of object, such as a media object, computer program, or food product to a person, organization, machine, or even an event. *Contradicted-by* could link a mathematical theorem to a counterexample or the campaign statement of a politician to a set of facts. *Continued-by* is a navigational link that applies primarily to media objects but could also be used to indicate the sequence of elements in a classification or the sequence of courses in a meal; *continued-by* has three arguments so that it can express the fact that an object can participate in many paths and the continuation object depends on the path. Finally, the relationship *has-narrower-term* illustrates that thesaurus relationships are no different from any other relationships in the system, such as *deals-with*, and can be used the same way in searching.

Other relationships are specific to an object type. The following examples are specific to media objects and data sets: *has-summary* leads from a long to a short form (a reader needing more detail than the summary provides could also pursue such a link the other way). *Has-prerequisite* serves a user who does not understand a media object and needs to acquire additional knowledge before reading it; *has-prerequisite* is also useful for constructing a didactically sound path. *Has-same-content-as* can be used by the system to select from among several media objects that say the same thing the one that best fits the user's language ability and cognitive style; to warn a user who is about to read something that it merely repeats what was just read; or, conversely, to suggest a different presentation if the user has trouble understanding.

Has-ingredients is used to model actual data about the ingredients of a food product. The structure is more complex to accommodate the complexity of the data. *Brownie has-ingredient Chocolate* does not tell the whole story. One also wants to know that chocolate has Rank 3 in order of predominance, amounts to 20 percent

121

based on total weight, 25 percent based on solid weight, and serves the purposes *improving taste* and *chewy consistency*. The reader who thinks this relationship too complex should consider the ingredients of a pudding. A pudding made with fluid milk might show milk as the first ingredient on a total weight basis, but a similar pudding made with dry milk shows milk as the third or fourth ingredient, even though the puddings' composition is essentially the same—as could be seen from ingredient information on a dry weight basis. Statements formed with the relationship *has-ingredient* serve many functions: they inform the user about the ingredients of a food, they let the user find all foods containing chocolate chips or chocolate in any form, they let the user find all ingredients that have been used to achieve chewy consistency, etc.

An information system should also include user models; that is, data about users (persons, organizations, machines), their characteristics, and needs. User model data are often kept separately, but the system envisioned here includes them in the information base just like any other data, represented just the same way through relationships. So are data about the relationship between users and other objects. For example, *readable-by* refers to the physical readability of a media object by "object1" (a person or device, such as a computer) with the aid of "object2" (another device, such as a microform reader or a disk drive of a certain kind). *Processable-by* refers to the capability of the agent object to do something with the data read, e.g., a file being processable by a program or a text being processable by a person (I can physically "read" a printed Russian text, but I cannot process it). *Processable-by* may link a media object to a computer program, which in turn can participate in many other relationships. It is therefore not advisable to create a class of readers separate from objects as suggested in Tompa 1989.

Further object types and relationship types could be introduced to support versioning [Halasz 1988, p. 847] and to integrate data on work status, schedules, and projects of individual users. This idea could be extended to support collaborative work [Halasz 1988, p. 848].

The groupings in Table 5 should not be construed too narrowly. For example, *has-prerequisite* could be used to link computer programs A and B where running A requires B. *Written-in* clearly applies to both documents and computer programs. *Has-ingredient* applies to drugs as well.

Object types and relationship types together specify what kind of data can be expressed in a hypermedia base/database. They define the conceptual schema.

On the surface, a statement about the ingredient of a food product may look quite different from a link between two text objects, one supporting the other or one continuing the other. But they can be expressed in exactly the same format, and there are many advantages in doing so. On the shell level an information structure management system combines the selective and inferential power of database and expert systems with the interface and navigational power of hypermedia systems. On the content level, one integrated conceptual schema is more parsimonious than several conceptual schemata, since many relationship types apply across content areas.

More important, an integrated schema allows for useful relationships that would

Table 5. Relationship Types (Link Types)

General relationship types	Relationship types applying primarily to media objects
produced-by	
has-target-audience	*has-prerequisite*
supported-by	*has-summary*
contradicted-by	*has-same-content-as*
praised-by	*is-simplified-from*
criticized-by	*is-later-version-of*
object deals-with subject/	*is-written-in*
subject dealt-with-in object	
	describes (the reciprocal of *described-in*)
described-in	
	illustrates
includes	
	Relationships on food products
has-special-case	
	Food product *has-ingredient*
has-narrower-term	[Food product, rank, total %, solids %,
	[purpose list]]
[object, path] *continued-by* object	
	Food product *underwent-process*
	[Process, equipment, temperature,
Relationships on issues, objectives, strategies	duration, place/stage, sequence no.,
	[purpose list]]
media object *helpful-for* problem/	
problem *help-in* media object	Food product *has-constituent*
	[ChemSubst, rank, total %, solids %]
circumstance *causes* issue	
	Relationships for user model
objective *addresses* issue	
	person *has-interest* [subject, intensity]
objective *addresses* cause	
	person *has-knowledge-of* [subject, depth]
strategy *aims-at* objective	
	person *reads-language* [language, fluency]
strategy *assigned-to* organization	
	media object *readable-by* [object1, object2]
	media object *processable-by* object

be much more cumbersome in separate systems. For example, a user could find all food products having 20 percent or more chocolate, select one of them, follow a link to a text describing it or a picture depicting it; or the user could define the neighborhood of all media objects that describe the selected product or any of its ingredients. Alternatively, a user could apply a program that computes the nutrient content data of a prepared food product based on ingredient data and nutrient data for simple food products and then pick a nutrient on the list and retrieve all media objects that discuss the importance of this nutrient.

Neighborhoods and Queries

A neighborhood (also called region or cluster) is any group of objects together with the relationships that exist among the members of the group (Table 6). More generally, a neighborhood can in turn contain neighborhoods. A neighborhood is not merely a set since it also includes the relationships between the members. The term was chosen in keeping with the spatial and navigation metaphor often associated with hypermedia systems.

Usually the members of a neighborhood are selected based on their relationship with one or more other objects. In the simplest case, a neighborhood consists of all objects that can be reached from a starting object through links of a given type—the neighbors of the starting object. For example, the neighborhood might consist of all documents criticizing a given document, which is assembled by starting from the given documents and following links of the type *criticized-by*.

Table 6. Neighborhoods

A neighborhood is any group of objects or neighborhoods, particularly a group identified through links with one or more other objects, together with the relationships that exist among the members of the group.

Examples

All documents *criticizing* a given document, together with the relationships among them (such as one of these documents citing another)

A search to identify this neighborhood would start from the given document and follow links of the type *criticized-by* to find the critical documents.

All pictures *illustrating* a given paragraph

All objects *dealing-with* a given subject

All objects *produced-by* a given person, e.g. all music pieces composed by
Gershwin

All persons *producing* a given document

All food products *having-ingredient* a given food product

All food products that occur in a *has-ingredients* statement in the ingredient role associated with the purpose chewy consistency

All objects selected by the user during the course of a search for later examination and processing

---

A direct association exists between neighborhood and search. A query leads to a neighborhood, and to each neighborhood corresponds a query. (Halasz [1988, p. 846] talks about "virtual composite nodes" that result from invoking a query and notes immediately that "it would make sense for the query language used for virtual structure descriptions to be the same as the query language used for searches and interface filters." The model suggested here does just that.) Including queries as objects allows for any kind of statements one wishes to make about a query and thus about the neighborhood defined by the query [Halasz 1988, p. 846].

So far we have discussed the most common way of defining a neighborhood—definition through a query. A neighborhood can also be defined by enumerating its elements. For example, a user may in the course of a search mark objects that appear interesting for inclusion in a retrieval list in order to come back to them later for examination and processing. In doing so, the user updates the information base: in effect adding an object that is the basis for the neighborhood being defined and links from that object to all the members of the neighborhood. The newly created object could be a document node with *includes* relationships to the objects being marked, or a subject node with *dealt-with-in* relationships to the objects being marked. Thus, the neighborhood created can subsequently be defined by a query.

Neighborhoods Based on Hierarchical Relationships

Neighborhoods based on hierarchical relationships between objects are an important special class (Table 7). Following a hierarchical relationship downward yields an *offspring neighborhood*. The first three examples are from a bibliographic context and need no further explanation. An offspring neighborhood of media objects lets the searcher specify a scope for a Boolean search condition; he can require two subject descriptors to co-occur in the same paragraph (a very strict condition) or in an entire book (the neighborhood of all paragraphs included in the book, a much looser condition). An offspring neighborhood of concepts implements inclusive searching, searching for a descriptor and all its narrower descriptors (easily done in MEDLINE and other systems using commands such as *explode* or *cascade* or, in DIALOG, an ! following the descriptor). An offspring neighborhood of assertions supports a thorough search for evidence disproving the assertion. For example, in a search for counterexamples that disprove a general mathematical theorem, one should look for counterexamples for any of the special cases as well. These examples illustrate the power of the general concept of an offspring neighborhood—one formalism handles three seemingly different situations.

The converse of an offspring neighborhood is an *ancestor neighborhood*. The ancestor neighborhood for a paragraph consists of all the media objects in which the paragraph is included: the article containing the paragraph, the journal issue containing the article, the journal volume to which the issue belongs, and finally the journal itself.

Ancestor neighborhoods provide an approach to hierarchical inheritance and are thus important in searching. As an example, consider a medical journal article deal-

---

Table 7. Neighborhoods Based on Hierarchical Relationships

Offspring neighborhoods

    A book and its chapters (one level down)

    A book and its chapters, subchapters, sections, paragraphs (all the way down)

    A journal and its volumes
    A journal volume and its issues
    A journal issue and its articles.

    A concept and its narrower concepts.
        Example: {cognitive processes (the top concept), apperception, cognitive mapping, serial ordering, associative processes, mental concentration, ideation, thinking}

    An assertion and its special cases

Ancestor neighborhood

    A paragraph in a journal article, the article itself, the issue, the volume, the journal

        Paragraph linked to treatment
        Journal article linked to hepatitis and to a person as author
        Journal linked to written for the general public

---

ing with hepatitis. A paragraph in that article might say "Five patients were treated with . . . . Three patients responded to treatment in 7 days . . . ." A user searching for paragraphs on the treatment of hepatitis would not find the paragraph in question unless the system adds the hepatitis link from the document as a whole. For another example assume that the user is interested in an explanation of AIDS for the educated lay person. The system must select articles based not only on their subject matter but also on the intended audience. The intended audience can often be inferred from the journal in which an article appears. All articles in *Scientific American*, for example, should inherit the attribute *has-target-audience* "Educated lay person."

An object that heads an offspring neighborhood plays an important role as an *organizing node*. Offspring neighborhoods together with the neighborhood links to be discussed below provide a "composition mechanism" postulated in Halasz [1988, p. 843] as "a way of representing and dealing with groups of nodes and links as unique entities, separate from their components."

Links associated with an organizing node should be assigned one of two roles. **Non-inheriting links** pertain to the organizing node as such or to the totality of the nodes under it; for example, "Scientific American *deals-with* All of science" is a non-inheriting link since it applies only to the journal (the totality of articles) as a whole, not to individual articles. **Inheriting links** are introduced as a space-saving device. An inheriting link applies to every object in an offspring neighborhood formed along a given relationship; for example, "Scientific American *has-target-audience* Educated lay person" is an inheriting link since it applies to every article *included-in* that journal. It would be a waste of storage space to make such a link for every article; it is much more efficient to make an inheriting link once with the top node for *Scientific American* itself and then apply it to every article through a mechanism of hierarchical inheritance (as discussed later).

Schema Neighborhoods

Schema neighborhoods are useful for the creation and display of structured documents, such as legal cases [Halasz 1988, p. 843]. As the example in Table 8 shows, a schema neighborhood consists of a head object of a given type (in the example the type "legal case") together with nodes related in one of a given number of ways. A system can support the creation of structured documents by asking the user for the type of document to be created and then displaying the appropriate schema template. It can support the display of structured documents by showing the schema outline when the head node is displayed. The relationship to frames need hardly be mentioned.

Schema neighborhoods share many characteristics of offspring neighborhoods as discussed in the previous section.

Table 8. Schema Neighborhood Example: Legal Cases

Head node for the legal case as a whole

The schema neighborhood for a legal case includes nodes related in one of the following ways:

      Legal case *deals-with-facts* object

      Legal case *deals-with-issues* object

      Legal case *reached-decision* object

      Legal case *gives-rationale* object

Each of these four objects is the head of an offspring neighborhood containing nodes for the individual facts, the individual issues, the elements of the decision, and the elements of the rationale, respectively.

Neighborhood Links

Neighborhoods give rise to neighborhood links, which are essential in searching. Neighborhood links are defined in terms of links between atomic objects as follows (Table 9): A *from-neighborhood link* is any link from any object in that neighborhood. As an example, consider a search for all text objects criticizing a given document; such text objects can be reached from the document following *criticized-by* links. The search should find text objects criticizing the document as a whole **or any part of it**; thus, it should start from the offspring neighborhood arising from the top document node and follow the relationship *criticized-by* from any object in that offspring neighborhood. A *to-neighborhood link* is any link that ends up in an object in that neighborhood. There can be a neighborhood on one or on both ends of a link. A good example of a link from a neighborhood to a neighborhood is a journal-to-journal citation link that exists whenever any article from journal A cites any article from journal B.

Table 9. Neighborhood Links

From-neighborhood link

    A link from any object of neighborhood N

To-neighborhood link

    A link to any object of neighborhood N

Paths and Scripts

Often a user is better served by reading a logically arranged sequence of media objects than by hopping from node to node. A *stored path* prepared by an editor allows just that. Such a path corresponds to a traditional article or book. When a computer program is divided into a number of pieces, some orderings are necessary for compilation and additional orderings are helpful for understanding by a human reader. A user may also want to preserve his or her own path through the hypermedia base.

Technically a **path** is a neighborhood in which *continued-by* relationships induce a linear ordering of all the objects in the neighborhood. There is one object for the path as a whole. This object contains the path name and may be linked to the creator, target audience, subject descriptors, etc. It must also be linked to the first actual object of the path. This object leads in turn to the second object through a *continued-by* relationship which is qualified by the path to which it belongs (see above, Table 5). Alternatively, the objects on a path could all be linked directly to the path node by an *includes-with-sequence* relationship. This relationship is a special case of *includes*; it imposes a sequence on the included objects. This detail of implementation is of little concern to the user.

An object on a path can in turn be a path. Thus a book can be represented as a path of chapters, each chapter as a path of sections, and each section as a path of paragraphs and figures.

A **script** is an object that contains instructions that "orchestrate the display of other [objects]" [Halasz 88, p. 846]. A script guides the user through an information base, possibly in a very prescriptive manner. For example, a script might organize a programmed instruction sequence that draws on a hypermedia base. As the script is run, information is presented, the user is asked questions, and the next item of information presented is chosen based on the answers. Another script might have instructions for putting together a document using text objects, retrieving data and calling a program to represent these data graphically, retrieving other data and applying a natural language generator, etc. Such a script could be called a *virtual document*.

The foregoing discussion has described the basic elements of an information structure. The remaining sections under this main heading add elements that make for a richer structure to express still more types of information or knowledge and to do more powerful searches.

Connections

A **connection** is composed of several links chained together. A link consists of only one step, but the searcher who wants to take several steps at a time can do so using a connection. For example, the chain

[Greek vase *has-instance* Object-1, Object-1 *depicted-in* Slide-567]

would lead from the descriptor Greek vase to a slide depicting one. The **connection type** in this case is

[Subject *has-instance* Object, Object *depicted* in Slide].

A connection type is defined as a chain of links, each link belonging to a specified link type or a specified neighborhood of link types. To preserve generality, we must broaden the definition of connection to include "chains" consisting of a single link. A connection consisting of only one link is a **direct connection**, a connection consisting of two or more links an **indirect connection**. It is often convenient to use a connection in the definition of a still more complex connection; thus on the most general level connection is defined recursively as a link or a chain of connections. Examples of direct and indirect connections are given in Table 10.

An indirect connection can be identified by a name, and that name can then be used just like a simple link, except that the system—transparent to the user—follows several steps. This procedure is equivalent to rules in Prolog: The connection name is the head of the rule, the individual links in the connection are the conditions.

Often an indirect connection has a meaning equivalent to a direct connection. For example, the indirect connection

[Person P *affiliated-with* Organization O, Organization O *has-phone* Phone number N]

is for most purposes equivalent to

Person P *has-phone* Phone number N

A good system should tell the user who employs a direct connection about

---

Table 10. Connections and Relationships Between Them

Sample connection type

[Subject *has-instance* Object, Object *depicted-in* Slide]

Direct and indirect connections

A **direct connection** connects two objects through a single link

    Examples

    Connection 1: [Media object A *commented-by* Media object B]

    Connection 2: [Media object A *criticized-by* Media object B]

    Connection 3: [Person P *has-phone* Phone number N]

An **indirect connection** connects two objects through two or more links via intermediate objects

    Examples

    Connection 4: [Media object A *proposes* Theory T,
                      Theory T *commented-by* Media object B]

    Connection 5: [Media object A *proposes* Theory T,
                      Theory T *criticized-by* Media object B]

    Connection 6: [Person P *affiliated-with* Organization O,
                      Organization O *has-phone* Phone number N]

Relationships between connections

    Connection 2 *isa* Connection 1

    Connection 5 *isa* Connection 4

    Connection 4 *equivalent-to* Connection 1

    Connection 5 *equivalent-to* Connection 2

---

equivalent indirect connections, which offer added possibilities for reaching relevant objects. To do so, the system must know such equivalences and other relationships between connection types. These relationships can then be used to create neighborhoods of connection types; we shall call a neighborhood of connections types a **bundle**. The user may choose to have the system do that transparently.

Complex connections (Table 11). The definition of some quite natural connection types requires more than just stringing link types together. For example, when looking for all objects that are hierarchically below (offspring) or above (ancestors) the system must follow a connection consisting of an unknown number of *includes* or *included-in* links. In a language such as Prolog one would use a recursive rule to handle this situation. In our context it should be possible to specify in the definition of a connection that a certain link could occur 1, 2, 3, or n times, where n could be a definite maximum or any number. The link could be defined by a single link type, such as includes. Or it could be defined less restrictively by a bundle of link types. Finally one could relax the restriction to the universal link type (any link type will do) and thus define a connection of a given length regardless of link type or, if the length is arbitrary, any connection at all.

Table 11. Complex Connections

[Subject-1 *RT (has-Related-Term)* Subject-2 (2)]

> Starting from Subject-1, this connection leads to other subjects that are one or two steps away following the *RT* relationship.

Example:

learning *RT* cognitive processes, cognitive processes *RT* **intelligence**

[Subject-1 *includes* Subject-2 (*)]

> Starting from Subject-1, this connection leads to other subjects included in Subject-1, an arbitrary number of levels down. (*Includes* is a very general relationship type; in a thesaurus one would use the designation NT for Narrower Term.)

Example:

cognition and memory *includes* **cognition**
    cognition *includes* **cognitive processes**
        cognitive processes *includes* **apperception**
        . . .
        cognitive processes *includes* **thinking**

[Object *deals-with* Subject-1, Subject-1 *includes* Subject-2 (*)]

> This finds all subjects connected with a known object, and then for each of these subjects all subjects included.

Data about Data, or Metadata

Much information and knowledge, especially expert knowledge, is information about information, or knowledge about knowledge. Belief strength and commentary are two examples dealt with in the following paragraph. On a more abstract level there is knowledge about the structure of knowledge, which is very important for searching. This topic is taken up beginning with the second paragraph below.

Statements about Statements or Statements as Objects

As discussed earlier, an information base consists of statements created by relating one or more objects using a given relationship type. Statements can themselves be objects that can participate in relationships. For example, if the statement *data set A supports assertion B* exists, and document C disputes that statement, the relationship of C is neither to the data set A alone nor to the assertion B alone but to the statement linking them. Similarly, if there is a statement *document A criticizes document B* and document C takes issue with that criticism, the proper linkage is not just from C to A or C to B but to the critique link between the two. Another important relationship type involving statements is the statement *is-believed-by* [object, strength], where the object could be a person or the system itself (for belief-strength indications representing the general consensus of the system builders).

Object Types and Relationship Types as Objects

An information base should be capable of expressing information (or knowledge) about object types and relationship types. The simplest way to meet this requirement is to treat object types and relationship types as objects about which statements can be made. Table 12 shows some relationships between object types (entity types). These are very useful for searching: The **isa** hierarchy lets the searcher specify *text document* as the target object type, and will find *journal articles*, *government reports*, etc. In effect the searcher defines an offspring neighborhood of object types. There are also relationship type neighborhoods: A user who starts from a paragraph and

---

Table 12. Relationships Between Object Types

Journal article *isa* Document

Government report *isa* Document

Text object *isa* Media object

Map *isa* Visual object

Visual object *isa* Audiovisual object

Sound object *isa* Audiovisual object

Audiovisual object *isa* Media object

City *is-part-of* address

---

wants to find paragraphs commenting on it should find all paragraphs linked by *commented-by*, *supported-by*, or *criticized-by*.

On a formal level, relationships between object types are no different from relationships between any other objects, and the system should treat them that way both internally and in the user interface. The operation to see the object types that fall under *document* should be no different from the operation to see all objects included in a given book or all concepts narrower than *system of government*.

Table 13 shows some relationships between relationship types (link types). The *commented-by* example already illustrated the use of a relationship type along with the relationship types under it (in the example *criticized-by* and *praised-by*). This example works also the other way around: A user who starts from a paragraph and wants to find all paragraphs criticizing it, should follow not only *criticized-by* but also *commented-by*, using an ancestor neighborhood of relationship types.

The broadest relationship type is the universal relation or link. Using the universal relation in indexing means that the link type is not specified. Using the universal relation in searching means that any link type is acceptable in the search.

Table 13. Relationships Between Relationship Types

Media object *criticized-by* Media object	*isa*	Media object *commented-by* Media object
Media object *praised-by* Media object	*isa*	Media object *commented-by* Media object
Person *is-author-of* Media object	*inverse-of*	Media object *has-author* Person

Universal relation or link as the top of the isa hierarchy

    In indexing: relationship type not specified

    In searching: any relationship type acceptable

Links as Program Calls

When a target object is not available it can often be generated. For example, a user may start from a given data set and look for pictures that represent it graphically. Suppose that no such picture is found but the system includes a graphics program that could generate one. The system must provide for a link from the relationship type *graphically-represented-by* to the graphics program, use that link to find the program, call the program with the data set as argument, and display the graphical representation on the screen. All this is transparent to the user, who searches for a graphical representation in the information base and finds it by following the appropriate link. Since there is in fact no such image stored, the image can be considered as a **virtual object** and the *graphically-represented-by* link as an **implied link to a virtual object** (implied because it is not stored explicitly but inferred from the type "data set" of the starting object).

Here is another example of a virtual object: If the relationship *other-language-version/French* does not find a text object, the system could invoke a translation program. Note that these programs are objects or neighborhoods in the information base; there might be a node for the overall program and hierarchically lower nodes for the program modules, other nodes for documentation, etc.

In a virtual link, the call to a program is implicit and transparent to the user. An explicit call to a program or function, either by entering the name or selecting from a menu, could also be seen as invoking a link.

Digression: Programming Environments as Information Structure Management Systems

Programming environments are now designed as **hypermedia/database systems** with exciting new possibilities. A brief sketch of a hypothetical system is included here to further illustrate the power of the unified systems approach.

A programming environment requires additional object and relationship types that open new ways of supporting the programmer (Table 14). When the programmer modifies a program, the system can display all calls to that program (particularly handy where the change involves the parameters needed in a program call). When the programmer modifies a variable definition or modifies a program part that changes a variable value, the system can display all programs using that variable. The most helpful part might be the linkage to documentation. Whenever a program part is changed, the system shows the corresponding parts of the documentation (both system documentation and user manual) or at least alerts the documentation editor.

The following example illustrates even tighter integration. Programs and user documentation often include the same value lists (for example, a list of relationship symbols allowed in a thesaurus). Rather than maintaining such a list in both places and risking inconsistency, one should be able to maintain just one list with appropriate format conversion for program or documentation use, respectively. The programmer can search the program (function) library for programs serving a given purpose and select from the retrieved list directly into that program. These are just a few examples of the support an information structure management system could provide to the programmer.

SEARCHING

Finally we come to the purpose of the whole elaborate structure—searching. Searching can be a very complex task, and this is reflected in our discussion. Much of this complexity, however, can be hidden from users who are willing to accept choices by the sys-

Table 14. Relationships Between Relationship Types

Additional object types

Variable

Instruction

Program segment, function (module, routine), program file are all covered by object type Program. A hierarchy is established by *includes*

Program type (with values source file, object file, and executable file)

Program purpose (such as sorting a list or extracting a substring)

Value list

Menu text

Additional relationship types

program *calls* program

program *serves* program purpose

program *defines* variable

program *changes* variable

program *uses* variable

program *affects* program (This relationship can often be inferred.)

media object *documents* program

program *includes* value-list

media object (part of documentation) *includes* value-list

tem, and the system can help those users designing their own searches by guiding them through the process and displaying menus of options where applicable. Complexity behind the scenes adds power and makes users' lives easier.

Definition of Search

A total search consists of one or more **search steps** (Table 15). For example, a user begins with a problem needing solution; the *help-in* link leads to media objects that can help in solving the problem. In a second step the user starts from one of these media objects and uses the *has-prerequisite* link to find media objects needed to acquire required background knowledge. The user selects one of these, finds it too difficult, and uses the *has-simplified-form* link to find a media object that is easier to read.

Table 15. Search: Definition

Total search = series of search steps, each leading

    from something known (for example, a known object or neighborhood)

      to something unknown but expected to be helpful (one or more helpful target objects or neighborhoods)

An object or neighborhood encountered may be helpful because it

    contains some of the information needed and/or

    serves as a stepping stone to other objects or neighborhoods.

To give another example, a user starts with a term and, using links among terms and concepts, finds the preferred term as well as broader and narrower concepts (a concept neighborhood). In the next step the user starts from this concept neighborhood and, using the *dealt-with-in* link, finds media objects.

Each search step starts from something that is known, often a **starting object** or **neighborhood**, and leads to one or more **target objects** or **neighborhoods** that, one hopes, contribute to the goal. Each object or neighborhood found along the way may contain some (or all) of the information needed and/or may serve as a steppingstone to further information, as the starting point in the next search step.

Searches differ in the method used to get from the known to the targets (Table 16). In the search examples given above, the method is navigation based on links (binary relationships). The user specifies a query by selecting an object on the screen as the starting object and selecting a type of link to follow. The same result could be achieved by formulating an explicit query. More complex cases, particularly searches using higher-order relationships, require an explicit query. The distinction between search as navigation and search as query-based retrieval [Halasz 1988, p. 841] is more a matter of perspective than of the basic nature of the search, and it is a matter of gradation rather than absolute difference (see the earlier section on multidimensional design space). The nature of the starting object (for example, media object vs. subject descriptor) may also play a role in determining the perspective. As the discussion and the examples in the next section make quite clear, the principle is always the same: The user starts from an object or neighborhood and, following a given link type, finds other objects or neighborhoods.

The typical hypermedia search uses the navigation metaphor. The starting object is the media object (a paragraph of text, a picture, etc.) currently on the screen, and the

---

Table 16. Types of Search

Search based on relationships between objects

    Search based on using relationships from one or more starting objects to identify target objects

        Two perspectives:

        Search as navigation (based on links, i.e. binary relationships)

        Search as query-based retrieval (based on relationships of any order)

    Similarity search based on relationships: Find objects with a neighborhood similar to a specified "query neighborhood". The query neighborhood can be specified as the neighborhood of a starting object

Search based on intrinsic properties of objects

---

user expects to find just one or, at most, just a few media objects to look at next. The media object found then becomes the starting point for the next search step, and so forth. A system might facilitate this process by showing on the screen a "map" of the links between objects, an outline being a special case. An alternative metaphor for the same transactions is the "bring" metaphor [Marchionini, personal communication].

The typical bibliographic search uses the query metaphor. The starting object is a subject descriptor or an author name, and a link type is often specified. The user won't be surprised finding 30, 50, or even 200 documents. A query can also be more complex, combining several starting objects and requiring that a target be reachable from all. Query formulations are also required for searching data represented through relationships with many arguments, for building elaborate inferences into a search, or for deriving new data through processing. Such query formulations can get quite complex.

Finally, a search can be based not on the links between objects or neighborhoods but on their content and internal structure. This is a truly different kind of search; it is always query-based.

The result of a search is a set of objects or neighborhoods. Thus the result is a neighborhood, and the query can be seen as the top node of that neighborhood.

The rest of this section on searching discusses various types and aspects of search, giving many examples to illustrate the power and generality of the approach. It first addresses navigation searches, moving from the simple (single-criterion searches starting from a single object), to the complex (combination searches with neighborhoods as targets and searching with hierarchical inheritance). It next introduces search based on the similarity of the neighborhood surrounding an object to a query neighborhood. Both of these search types are based on the relationships of objects or neighborhoods to the outside. Finally, we consider a quite different type of search—those based on internal properties, either the structural properties of a neighborhood derived from the configuration of relationships within the neighborhood, or internal properties of atomic objects. The remaining paragraphs introduce refinements. The systems described in Croft [1989] and Thompson [1989] illustrate several of the approaches that we discuss here in a general framework.

Specification of a Search Based on Relationships

This section discusses in detail how to specify a search based on using relationships from one or more starting objects to identify target objects. For ease of explanation the discussion uses the navigation metaphor, but the concepts apply more broadly to any search based on relationships.

In each search step the user must tell the system what is sought through a search specification (query formulation) consisting of the four elements shown in Table 17.

The **target object** or **neighborhood specification** expresses what the user wants to find—concepts, persons, media objects, offspring neighborhoods of media objects, ancestor neighborhoods of media objects, assertions, etc. Specifying neighborhoods as targets is significant in a Boolean AND search (see below).

As a **starting object or neighborhood** for the first search step the user may employ either a known object (a problem, a media object, a term, a person) or select an object from an initial menu. Alternatively, the user may enter a **starting object type**, such as **problem**, and in return be shown a menu of possible values to select from.

How to get from the starting object to the target objects wanted is specified in a **connection condition**, which consists of **one or more permissible connection types**. Each connection type might be assigned a weight and the weights **combined** in some way to result in a relevance weight for each target object or neighborhood. (When weights are used, neighborhoods become fuzzy sets. For now, lets consider only weights 0 and 1 and Boolean combinations.)

Often the user's purpose is achieved best by allowing any of a number of related connection types; this can be done by specifying a bundle of connection types. The bundle could mix direct and indirect connections. As an example, assume a user is looking at a given media object A and wants to find media objects in which A is criticized. Table 10 gives two connection types that lead to such objects, so the user should specify the connection condition as the bundle

---

Table 17. Elements of a Search Specification (Query Formulation)

The general type of target objects or neighborhoods

One or more search criteria, each consisting of

    A starting object or neighborhood

    A connection condition: the permissible connections from the starting object or neighborhood to the target objects or neighborhoods.

The format in which the target objects or neighborhoods found should be displayed.

{Connection 2 [Media object A *criticized-by* Media object B],
Connection 5 [Media object A *proposes* Theory T,
    Theory T *criticized-by* Media object B]}

For a more complete search, the user should also include the broader link type **commented-by** and specify the bundle

{Connection 1 [Media object A *commented-by* Media object B],
Connection 2 [Media object A *criticized-by* Media object B],
Connection 4 [Media object A *proposes* Theory T,
    Theory T *commented-by* Media object B],
Connection 5 [Media object A *proposes* Theory T,
    Theory T *criticized-by* Media object B]}.

Once target objects or neighborhoods are found they must be displayed using the **display format** given by the user. The display format specifies the information to be given for each target object or neighborhood. For example, the user may want to see just titles, or the author and the title, or the full text. Gathering this information may mean an implied search (transparent to the user). The format also specifies the arrangement of the objects or neighborhoods found. One possible format is a network based on a specified link type.

Searches differ in formal complexity: A search can use a single search criterion or a combination of search criteria. It can start from a single object or from a neighborhood. The targets specified can be single objects or they can themselves be neighborhoods. The following discussion expands on search criteria and their combination.

Single-criterion Search Starting from a Single Object

This is the simplest kind of search. In a single-criterion search, a single connection going into a target object is sufficient to select that object (Table 18).

In the examples in Table 18, atomic objects are specified as targets for selection. Sometimes one may want to retrieve whole neighborhoods, for example whole documents rather than single paragraphs (Table 19). Put differently, the whole document should be shown if any of its subordinate objects (sections, paragraphs) are found. In that case, the search targets can be specified as document offspring neighborhoods. The system finds any media object that fulfills the search criterion and then identifies the whole document to which it belongs. Thus specifying neighborhoods as targets of a single-criterion search affects not retrieval per se but what information is displayed once an object is found. This result is in contrast to combination searches where, as we shall see, it matters greatly for retrieval whether targets are atomic objects or neighborhoods.

Single-criterion Search Starting from a Neighborhood

A single-criterion search starting from a neighborhood can start from any object in the neighborhood (Table 20). It is an implied **OR** search, which leads to many more objects than a search starting from a single object in the neighborhood.

Table 18. Single-criterion Search Starting from a Single Object

Examples

      Starting object: A subject descriptor

      Target objects: Media objects

      Connection condition: *dealt-with-in*

      Starting object: A media object

      Target objects: Descriptors

      Connection condition: *deals-with*

      (The descriptors found might be useful as starting points in further search steps)

      Starting object: A building

      Target objects: Media objects

      Connection condition: {*depicted-in*, *discussed-in*}

      Starting object: A book   (only the top node for the book as a whole, excluding the nodes for parts of the book)

      Target objects: All objects of any type

      Connection condition: *Universal link* (any link type)

      Starting object: A book (top node only)

      Target objects: All objects of any type

      Connection condition: Any one-link or two-link connection

---

The first example is an inclusive subject search. The system first builds the offspring neighborhood consisting of a descriptor and all its narrower descriptors, and then starts from any descriptor in that neighborhood (rather than be limited to the one descriptor that heads that part of the hierarchy) to find media objects via the connection type *dealt-with-in*. This is an implied **OR**ing of the starting descriptors. In searching MEDLINE this can be achieved by prefixing a descriptor with **EXPLODE**.

Table 19. Single-criterion Search Starting from a Single Object with Neighborhoods as Targets

Example 1

A search for all whole documents that deal with a given descriptor, such as **Drug treatment**. The descriptor could be assigned to the document as a whole or to any section or paragraph included in it.

Starting object: Subject descriptor **Drug treatment**

Targets: Offspring neighborhoods from whole document nodes

Connection condition: *dealt-with-in*

A *dealt-with-in* link from the descriptor to any one element (section, paragraph) of a target neighborhood is sufficient.

Display: Show the whole document node, under it any sections that either has the descriptor assigned to it or includes a paragraph that has the descriptor assigned to it. Highlight the objects to which the descriptor is assigned.

Example 2

A search for all food products to which a person (or person class) is allergic. This search is complex because *allergic-to* statements can refer to food products, such as milk, or to a chemical substance, such as lactose, and because the offending allergen may be in any ingredient of the food. In other words, this search must find all food products that directly or indirectly contain something the person is allergic to.

Starting object: A person

Targets: Neighborhoods consisting of a food product, all food products reachable from it through *has-ingredient* links or chains of *has-ingredient* links, and all chemical substances reachable from any of these food products through *has-constituent* links

Connection condition: *allergic-to*

An *allergic-to* link from the person into any element of a target neighborhood is sufficient.

Display: Show the whole food and the offending ingredients or constituents.

The next example illustrates even better the usefulness of this method. An assertion is contradicted when any of its special cases is contradicted. So to see all objects that contradict an assertion, start from the assertion itself and find all objects linked via *contradicted-by*, but then also start from each of the special cases and follow the same link, that is, start from the offspring neighborhood. Finally, to find all the media objects linked to a given book, start from the book node itself but also from all the nodes dependent on the book node via a chain of *includes* links, i.e., an offspring neighborhood.

Table 20. Single-criterion Search Starting from a Neighborhood

Any of the objects in the starting neighborhood can serve as a starting point for the search, vastly increasing retrieval (implied Boolean OR).

Examples

Starting neighborhood:	A subject descriptor and all its narrower descriptors (MEDLINE EXPLODE) Example: {cognitive processes (the top concept), apperception, cognitive mapping, serial ordering, associative processes, mental concentration, ideation, thinking}
Target objects:	Media objects
Connection condition:	*dealt-with-in*
Starting neighborhood:	An assertion and all its special cases
Target objects:	All objects of any type
Connection condition:	*contradicted-by*
Starting neighborhood:	A book and all its chapters, sections, and paragraphs.
Target objects:	All objects of any type
Connection condition:	The universal link
Starting neighborhood:	A city and all locations in a 100 mile radius
Target objects:	All businesses
Connection criterion:	*is-location-of*

This type of search is very powerful since the user can define any neighborhood wanted. Some neighborhood types may be predefined in the system and be referred to by an operator (such as an offspring neighborhood). When the user applies this operator in the definition of the starting point of the search step, the system does a search, but it is transparent to the user. Other neighborhoods must be assembled through an explicit search step. The formalism is always the same.

Combination Search (Boolean AND or Weighted Search)

Single-criterion searches are simple but often overly general. More specific selection requires using two or more search criteria simultaneously—a combination search (Table 21). A straightforward combination search requires that a target object or neighborhood satisfy two or more search criteria simultaneously. A more complex weighted search computes a weight for each target object or neighborhood, considering the degree to which the target object or neighborhood satisfies each search criterion (which in turn might be a function of link strength) and the importance the user assigned to the search criteria.

The straightforward combination search can be restated as follows: For a target object to be selected in a combination search it must be reachable by two or more connections, normally coming from different starting objects. A combination search with direct connections each using a single link type amounts to straightforward Boolean AND. A combination search with connections of arbitrary length with arbitrary link types amounts to unconstrained spreading activation. The formalism described here lets the user specify anything in between, even adding a weighting scheme.

Table 21. Combination Search (Boolean AND) with Single Objects as Targets

Find all objects that satisfy two or more search criteria simultaneously.

Examples:

Starting from two descriptors, follow the link type *dealt-with-in*. The objects reached from **both** descriptors are the objects that deal with both.

Starting from two objects, follow the link type *deals-with* . The descriptors reached from **both** objects have been used in indexing both, as in the example already given in the text:

Document A *deals-with* alcohol, apperception, impairment, teenagers

Document B *deals-with* apperception, spatial, sex differences

Descriptor found: Apperception

Starting from two documents, find all documents *citing* both (co-citation).

Table 21 gives some simple and familiar examples. The first example is a plain Boolean AND search where the searcher specifies two descriptors and wants documents that deal with both. With the general search operator suggested here the searcher can turn this around: starting from two objects known to be relevant, she can find all descriptors that are used in indexing both objects. Those should be good candidate descriptors to find more relevant objects, much better than the descriptors used in indexing just a single relevant object.

Example:

Documents A and B are known to be relevant
Document A *deals-with* alcohol, apperception, impairment, teenagers
Document B *deals-with* apperception, spatial, sex differences

Clearly *apperception* is a much more plausible descriptor for finding more relevant documents than *alcohol* or *sex differences*.

In an extension of this model, the searcher specifies the set of all known relevant objects as the starting neighborhood and the system ranks descriptors by the number of documents they index.

Combination Searches with Neighborhoods as Targets

With combination searches it makes a big difference whether the targets are restricted to atomic objects or whether neighborhoods are admitted and, if so, what type of neighborhood (Table 22). Searching for neighborhoods uses to-neighborhood links or, more general, to-neighborhood connections. A to-neighborhood connection exists whenever there is a connection to any element of the neighborhood, greatly increasing the possibilities for simultaneous satisfaction of two search criteria. By proper definition of target objects or neighborhoods, the searcher can require, for example, that two terms co-occur in the same paragraph, in the same book chapter, or in an entire book, or in an entire journal issue, or in the neighborhood formed by a document and all documents it cites. The examples show that this concept has very broad application.

Table 22. Combination Search (Boolean AND) and Neighborhoods as Targets

A neighborhood satisfies a search criterion if any of its objects satisfies it.

Examples:

Starting from two descriptors,
find all neighborhoods of a given type indexed by both.

Examples for neighborhood types that make sense here:

Offspring neighborhoods starting from the top node for a whole document, such as book, journal article, or report, along the link type *includes*.

Ancestor neighborhoods of any media object along the link type *includes*.

The neighborhood consisting of a document and all the document it *cites*.

Specifying neighborhoods as the targets of a search requires definition of the kind of neighborhood desired, such as offspring neighborhoods starting at whole document nodes or citation neighborhoods starting at any media object following citation links one step (or two steps, or n steps). Fortunately, defining neighborhoods requires no new syntax, it is just the same as search specification; i.e., specify a type of starting object and the connection type(s) to be used to reach other elements of the neighborhood.

Today's hypertext systems do not allow the specification of neighborhoods as search targets. Thus it is not possible to conduct a whole-document level Boolean search even if Boolean searching is implemented. (For offspring neighborhoods, Frisse [1988] achieves a result somewhat similar to specifying neighborhoods as targets through a complex algorithm that draws on the previously computed relevance of the children when it computes the relevance of a parent node.) Many hypertext systems allow Boolean searching for an "index search" but not for a search using the typical hypertext links. In bibliographic ISAR systems the search level supported depends on the type of descriptor used. Subject descriptors assigned through explicit indexing are assigned to documents as a whole; thus a search for whole documents is supported only when one uses assigned subject descriptors. With text words, one can specify as the search target a whole document or a paragraph or a sentence.

The concept of a neighborhood type as search target is so fundamental and has such broad applications that two more examples are in order. The first example (Table 23) generalizes the idea of finding descriptors in common to two relevant objects and using them as descriptor candidates for further searching. The search now targets **descriptor neighborhoods** that can be reached from two relevant objects. In this

---

Table 23. Combination Search (Boolean AND) and Neighborhoods as Targets 2.

Starting from two objects, find offspring descriptor neighborhoods that can be reached from both through *deals-with*

Example:

Documents B and C are known to be relevant

Document B *deals-with* apperception, spatial, sex differences

Document C *deals-with* thinking, verbal, bilinguals

Even though there ar no descriptors in common, from both documents one can reach the descriptor neighborhood

{cognitive processes (the top descriptor), apperception, cognitive mapping, serial ordering, associative processes, mental concentration, ideation, thinking}

Any of the descriptors in that neighborhood are good candidate starting points for finding more relevant documents.

---

example, the system identifies broad descriptors in common to two objects even though each broad descriptor is represented by a different one of its narrower descriptors in each of the two documents. Such descriptors can be extremely useful for further search.

Hierarchical Inheritance through Ancestor Neighborhoods as Search Targets

The neighborhood consisting of a media object and all its ancestor objects is of particular interest (Table 24). A Boolean AND search is satisfied if a proper connection exists either to the lowest level object itself or to any of its ancestors through an inheriting link. This can be interpreted as hierarchical inheritance: for purposes of this search, the lowest level object inherits selected connections into any of its ancestors. The hierarchical inheritance is qualified by the specific hierarchical relationship (or bundle of relationships) specified.

Table 24. Hierarchical Inheritance through Specifying Ancestor Neighborhoods as the Search Targets.

D359 (the section **Drug treatment**) *included-in* D355 (the article **The spread of AIDS**)

D355 (the article) *included-in* D243 (Scientific American)

D243 (Scientific American) *has-target-audience* Educated lay person
or Educated lay person *is-target-audience-of* D243

D355 (the article) *deals-with* AIDS or AIDS *dealt-with-in* D355

D359 (the section) *deals-with* Drug treatment
or Drug treatment *dealt-with-in* D359

Query: Find using ancestor neighborhoods that meet the following criteria:

Reachable from **Educated lay person** through *is-target-audience-of* AND

Reachable from **AIDS** through *dealt-with-in* AND

Reachable from **Drug treatment** through *dealt-with-in*

Ancestor neighborhood:

{D359 (the section), D355 (the article), D243 (Scientific American)}

This neighborhood meets all three search criteria.

Assume *Scientific American* published an article "The Spread of AIDS," which includes a section on "Drug Treatment." Table 24 contains the relevant statements in a hypermedia base. The section should clearly be found in a search for material on the **Drug treatment of AIDS** that is written for the **Educated lay person**. A Boolean AND search based on ancestor neighborhoods will accomplish this. The ancestor neighborhood of the section consists of three elements, namely:

[D359 (the section), D355 (the article), D243 (Scientific American]

Any link into any of the elements is a neighborhood link into the ancestor neighborhood. Thus the ancestor neighborhood meets all three search criteria and is retrieved.

Search As Inference

In the broadest sense, any search is the inference "if an object meets the search criterion (or criteria), it should be found." One is more likely to speak of inference when a search uses multiple criteria or when it uses a chain of links or relationships, thus combining multiple pieces of data to arrive at a conclusion about retrieval. For examples see the indirect connections discussed previously under the heading "Connections." To give another example, one might introduce the connection

C *may-support* A defined as the indirect connection [C *criticizes* B, B *criticizes* A]

Looking at search as inference greatly enhances the power of hypermedia systems.

Search for Objects Based on the Similarity of Their Neighborhood to a Query Neighborhood

So far this section has treated searching as navigation, as reaching target objects from a starting point. But searching can also be seen as examining the surroundings of each target object to see whether it meets certain criteria (Table 25). For example, the user could instruct the system to check each media object to see whether it has a *deals-with* link to the subject descriptor *hypertext*. Or the user could specify a combination criterion, asking for links to two subject descriptors.

Table 25. Similarity Search

Starting from an object or a neighborhood, find objects or neighborhoods with similar surroundings or define a "query neighborhood and find objects or neighborhood whose surroundings are similar to this query neighborhood.

Two objects or neighborhoods have similar surroundings if the patterns of links emanating from them are similar. This could be restricted to types of links, for example

Two media objects are similar if they share many subject descriptors.

Two media objects are similar if they share many cited media objects.

Two assertions are similar if contradicted by many of the same statements.

The search described is a simple exact-match search for the presence of two descriptors. A more complex search can compute a degree of match based on the presence of a fraction of the descriptors wanted. Even better, it could give some credit for related descriptors, where descriptor relationships can be given explicitly or inferred by similarity as described below.

The criteria can include all types of direct or indirect connections.

The user can also pick a known relevant object and request target objects that have similar surroundings with respect to one or more connection types. An example is objects that are indexed by the same descriptors as the known relevant object or at least show considerable overlap. Conversely, two descriptors are similar if the sets of objects linked to each show considerable overlap.

The possibilities for formulating similarity criteria are limitless. The user specifies the object to serve as the standard of comparison and the type(s) of connection(s) to be used in assembling the surroundings of the object. If the user simply specifies "similar," the system could use all direct connections to determine similarity. An important special case is a search for all objects that occur in a neighborhood of given structure in a given role.

The objects similar to a given object form a **similarity neighborhood**. Such a neighborhood could be used in any of the ways described above. For example the user could specify a Boolean AND search targeting similarity neighborhoods; that is, the user would instruct the system to select similarity neighborhoods that can be reached from both starting objects. However, such a search would require very extensive computation.

Search Based on Internal Properties

So far we have discussed searching based on the relationships or links between objects or neighborhoods. But selection of objects or neighborhoods can also be based on their internal structure or other properties. For example, free-text searching uses operators such as ADJacent, which requires that two words occur next to each other in the text (or no more than n words apart), or an operator requiring two words occurring in the same sentence. Free-text search systems that deal only with whole documents also provide an operator requiring that two words occur in the same paragraph. A hypermedia system that deals with paragraphs as objects does not need such a paragraph operator, since the search can just specify paragraphs as target objects. With parsers and computing resources both improving, we will soon see operators that allow the specification of syntax-based dependencies of words in text.

One can think of analogous examples in searching for pictures and for pieces of music. Assume a system that stores digitized images and that includes a picture recognition program capable of identifying subjects depicted, such as buildings and persons. One could than search for all pictures that depict a person inside a building. Similarly, with digitally stored music one could search for all pieces of music that follow the form abba.

The search targets can be atomic objects. Determining the internal structure of atomic objects requires programs, such as parsers, that can read the objects and process them to recognize the structure. [Halasz 1988, p. 842, calls this "content search."] The search targets can also be neighborhoods. Determining the internal structure of a neighborhood requires a structure-recognition program that processes the links within the neighborhood. [Halasz 1988, p. 842, calls this "structure search."] Note that this is quite different from retrieval based on external links that emanate from or enter a neighborhood.

An internal property search can use exact match or a similarity criterion.

Limiting the Search

A user may limit the search to a neighborhood. That is, search steps executed after the limit is put into effect return only objects belonging to that neighborhood. For example, in bibliographic systems it is quite common to limit by year of publication. Or a user, having finished his search of the entire system, may now want to limit the search to the neighborhood of objects selected for further examination. Or the user may want to limit the search to objects included in a given handbook or journal, in other words, to an offspring neighborhood.

A special case of limiting is the nesting of hypermedia bases: allow an entire hypermedia base B as a node in a superordinate hypermedia base A. When the user of A selects node B, he or she is placed into B for search (navigation) in B only. The subordinate hypermedia base B can be seen as a neighborhood in the superordinate hypermedia base A. In this approach, the neighborhood to which the search is limited is predefined; in limiting, in general, the user can define the neighborhood. Also with nested hypermedia bases, the nodes in B are not linked to nodes in A; such links might be quite useful for nonlimited searches.

Refinements of Search

The search methods considered so far do not give the objects or neighborhoods found in one search step in order of decreasing relevance, nor do they support the user in traversing the information base in a logical sequence. While some users might prefer to find their own way, others might welcome at least suggestions for a sensible path. A user could still decide whether to follow the suggested path and could detour from or completely leave the suggested path at any time.

Rank ordering the objects found by expected relevance requires computing a value for expected relevance as discussed previously (see Table 25).

Creating a logical path is much more difficult. Using a stored path or script is, of course, one solution. But a stored path is usually not optimally adapted to the needs of the individual user. The ideal system would consider the user's present state of knowledge, the state of knowledge required to solve the problem at hand, and then construct a path that leads from here to there. In an education context, the system could produce a "textbook" tailored to the background and learning objectives of an individual student. A discussion of how this could be done is well beyond the scope of this chapter, so a few hints must suffice. The system needs to rely on *has-prerequisite* links to select a media object that transmits the information needed without requiring knowledge the user does not have, or, if that is not possible, include in the constructed path a media object that provides the prerequisite knowledge. It must also consider something more subtle: the reading of one media object may train the user in a way of thinking that is helpful in understanding another media object. This kind of relationship establishes a basis for sequencing. The system must consider the user's background when choosing examples. The system must adapt to the user's cognitive style; for example, does the user learn best when examples are given first and general principles stated afterward or when the general principles come first and are then illustrated by examples.

Assembling Result Neighborhoods During Search

As the search proceeds, the user may mark items for further examination or for use in a given purpose, creating one or more **result neighborhoods**. This requires no new functionality. The user can simply define new objects corresponding to the result neighborhoods to be created (we may call these objects markers) and establish a link between a found object displayed on the screen and the appropriate marker(s). The system should facilitate this process. For example, the system could create temporary markers 1 - 9, which would exist until the end of the session or until a completely new search is started. Before the system erases a temporary marker and the associated links, it would give the user a chance to give it a permanent name. A marker may, in effect, become a descriptor that would be useful to other users as well.

INDEXING

The relationships or links used in searching do not come out of nowhere—somebody or some system must first establish them. There are explicit relationships that are stored as such in the information base, and implied (computed) relationships derived through inference and similarity computation, or by the determination of an optimal reading sequence. Once derived, implied relationships can be added to the database as explicit relationships.

Indexing in the general sense is the creation of any type of explicit relationship. Such relationships can be created by the creator of an object, by specially appointed editors/indexers, and by users.

Explicit relationships include many relationship types that in some systems might be called computed relationships, especially relationships based on information provided by the author of a document. For example, the author of a document creates *includes* and corresponding *continued-by* links from the top node for the entire document to the sections, and from each section to paragraphs and figures. In a printed document these links are expressed by the arrangement, and for inclusion in a hypermedia base the representation must be changed, but they are nevertheless explicit links. The author may also create a *has-abstract* link to another text object. By her choice of words, the author creates links from the document or any of its parts to the word, and conversely from the word to the document or a part of it.

Further relationships can be created by editors or indexers. In the document example, an indexer may create several links between subject descriptors and the document as a whole as well as links between subject descriptors and individual document sections or even individual paragraphs. An indexer may introduce (create) subject descriptors and create links between them. In that example, the indexer is perhaps better called "thesaurus builder." Many might not even call establishing links between subject descriptors "indexing," but it is establishing links and requires intellectual decisions. Having stated the fundamental sameness one should add that there are also differences: establishing links between descriptors does involve a different link type, requires a different type of thinking, and has more far-reaching, system-wide consequences.

Indexing with Hierarchical Inheritance

When dealing with an object, such as a document, that has multiple objects under it, the indexer must decide at which hierarchical level a relationship should be made. The indexer should use a principle known from semantic networks and frame systems: if a relationship applies to all or almost all subordinate objects, establish the relationship at the superordinate object as an inheriting relationship; if the relationship applies only to a specific object, establish the relationship only for the specific object. This principle ensures parsimonious indexing with no ill effect for searching, provided hierarchical inheritance is applied as, for example, by specifying ancestor neighborhoods as targets as described previously in the section "Hierarchical Inheritance Through Ancestor Neighborhoods as Search Targets."

Table 26 gives examples. The first example deals with a hierarchy of media objects: a journal, an article, a section of the article. A statement that holds for all objects (e.g., all sections) under a superordinate object (e.g., an article) should be made at the superordinate level.

The second example deals with a hierarchy of descriptors. It restates a well-known indexing rule: rather than assigning several specific descriptors that are all children of the same broader descriptor, assign the one parent descriptor that includes them all.

The third example also deals with a hierarchy of descriptors, this time in a thesaurus building context. Rather than giving the same RT (Related Term) relationship for several specific descriptors that are all children of the same broader descriptor, give the RT relationship for the one parent descriptor that includes them all.

REVIEW: THE FUNCTIONS OF OBJECTS AND RELATIONSHIPS

This section returns to this chapter's major theme—the essential unity of database systems, expert systems, ISAR systems, and hypermedia systems—through a discussion of the various functions that objects (entities) and relationships (links) play in a unified system.

151

Table 26. Indexing with Hierarchical Inheritance

Example 1:   Media object hierarchy

Indexing the section **Drug treatment** (D359) in the article **The spread of AIDS** (D355) in Scientific American (D243). Previous indexing resulted in the following statements:

D359 (the section) *included-in* D355 (the article)

D355 (the article) *included-in* D243 (Scientific American)

D243 (Scientific American) *has-target-audience* Educated lay person

D355 (the article) *deals-with* AIDS

The following statements are true for the section (D359): *deals-with* Drug treatment, *deals-with* AIDS, *has-target-audience* Educated lay person. However, only the first is unique to the section, the other two are already included for superordinate objects. Thus there is no need to repeat them; only one statement needs to be added:

D359 (the section) *deals-with* Drug treatment

Example 2:   Considering the descriptor hierarchy in descriptor assignment

If   Document D *deals-with* apperception

      Document D *deals-with* cognitive mapping

      Document D *deals-with* . . . (serial ordering, associative processes, mental concentration, ideation, thinking)

are all true, the indexer should assign the broader descriptor:

      Document D *deals-with* cognitive processes

conversely, cognitive processes *dealt-with-in* Document D is true.

This relationship inherits down to apperception, cognitive mapping etc. Thus,

      thinking *dealt-with-in* Document D   is true through inheritance.

A complete search for thinking should start from the ancestor neighborhood.

      {thinking, cognitive processes, cognition and memory (a still broader descriptor)}

Example 3:   Considering the descriptor hierarchy in establishing thesaurus relationships

      Each of the descriptors under cognitive processes is related to intelligence. But instead of eight relationships apperception *has-Related-Term* intelligence, cognitive mapping *has-Related-Term* intelligence, etc., we should establish just one inheriting relationship

      cognitive processes *has-Related_Term* intelligence

The Functions of Objects

Objects have five major functions (Table 27):

1. Objects participate in relationships that model actual data. One can make a

statement about chocolate chip cookies only if *chocolate chip cookie* is an object in the information base.

2. Objects, specifically media objects, represent data for assimilation by users. A paragraph, a figure, a tone document are meant to transmit information or, more generally, to enter the users cognitive or affective sphere.

Table 27. Functions of Objects (Entities, Nodes)

An object can serve one or more of these functions

Participate in relationships that model actual data

Represent data for assimilation by users (media objects)

Serve as an access point that leads to other objects

    Example: Subject descriptor
    But: Entry for subject descriptor also useful in itself, as in a dictionary.

Provide a focus for the organization of the database

Serve as focal points for relationships that pertain to all elements of a neighborhood (inheriting relationships)

---

3. An object may serve as a starting point or a query element to access other objects. Some objects are introduced primarily for the access function; for example, concepts to be used as subject descriptors are introduced not as objects of interest in their own right but to serve as access points for retrieval of media objects, software objects, organizations, persons, food products, or whatever. However, concepts may also be useful in themselves—their hierarchical relationship may convey information and their definitions may be of interest. The thesaural relationships given for a subject descriptor (Broader Term, Narrower Term, Related Term) may also contribute to the definition and thus help a user whose final information need is clarification of the meaning of a term. This illustrates a general point: an object can, and often does, serve several functions simultaneously.

4. Objects provide a focus for the organization of the information base. This function is very important for hypermedia bases. The head of an offspring neighborhood, such as the top node for a book leading to all the chapters, or a path object serve this function.

5. Objects serve as focal points for relationships that pertain to all elements of a neighborhood (inheriting relationships). For example, the relationship "Scientific American *has-target-audience* Educated lay person" is introduced with the idea that it holds for all the articles *included-in* any *Scientific American* (hierarchical inheritance).

Functions of Relationships (Links)

Relationships have two functions (Table 28). One function is to model actual data; the *has-ingredient* relationship for food products is an example. Another example is data set *supports* assertion. That is a factual statement made possible by including the relationship type *supports* in the conceptual schema.

The second function of relationships or links is to point to other objects. This function is used in retrieval and navigation. Relationships established primarily to

---

Table 28. Functions of Relationships (Links)

Modeling actual data

    Examples:

    Food product 1 *has-ingredient* [food product 2, ...]

    Data set *supports* assertion

Pointing to other objects

    Most links

    Examples for links established primarily as pointers:

    Media object *includes* media object

    Media object *continues* media object

    Concept *dealt-with-in* media object

---

model actual data are often used as pointers as well. If, on the one hand, I want to know the ingredients of chocolate chip cookies, I use the relationship *has-ingredient* for its substantive information value. If, on the other hand, I want to find all foods containing chocolate chips, I use the same relationship as a pointer for retrieval purposes. Links between media objects are established primarily for their pointer value. Media object 1 *includes* media object 2 does make a statement about these two objects, but the link's main purpose is to point the reader of media object 1 to media object 2.

The observation that objects (entities) and relationships (links) serve multiple functions, some more familiar in the database world and some more familiar in the hypermedia world, underscores the advantage of the unified view presented here. It shows that existing systems set up primarily to store factual data can be navigated (provided proper software support) and existing systems set up primarily for navigation can be used to look up facts.

DESIGN ISSUES

The implementation of a general information structure management system that would combine the functionality of a database management system, an expert system shell, a retrieval package (including text retrieval) and a hypermedia shell, is a tall order. Such a system must solve two major design problems: provide a powerful search engine that can implement the search options discussed, many of which require intensive processing; and provide an interface that facilitates using the powerful general search operators. The system must also provide a basic set of object types and relationship types with proper semantics for processing built into the system and allow the user to define additional object types and relationship types with their semantics. Table 29 lists these design issues.

A few remarks on the user interface must suffice here. For navigation-type searches, the user need only specify the starting object (usually an object on the screen) and select the link type from a menu associated with the starting object. The search targets are often implied by the link type. Where appropriate, the system should remind the user of the ability to specify an offspring neighborhood as a starting point to broaden retrieval, and it should guide the user in the proper choice of targets (atomic objects or neighborhoods). The system should also suggest related link types or connection types to follow.

Table 29. Design Issues

Provide a powerful search engine that can implement the search options discussed, many of which require intensive processing

Provide flexible user interface

> General search operator syntax that work for all types of objects and relationships.

> Simplified search operator syntax for frequent special cases with preset values and default values.

> Menu-driven search specification with preset values and default values.

Certain types of neighborhoods (offspring, ancestor) available through proper notation without explicit search specification

Neighborhood schemas definable by users

Scripts

A basic set of object types and relationship types with proper semantics for processing built into the system

User-definable object types and relationship types with user-definable semantics

---

The user should be able to refer to frequently used neighborhood types—offspring neighborhoods, ancestor neighborhoods, schema neighborhoods—by a simple symbol without formulating a query. If an offspring neighborhood could be formed along any of several hierarchical link types, the system should ask.

Once the system constructed a query based on the user's answers, it should ask whether the user wants to save it under a name for later use and possibly editing. The system also needs a syntax for query formulation by more experienced users. This language should incorporate all the capabilities of SQL and of a language such as Prolog but not be limited to those.

As with a database management system, specific applications need their own definitions with additional object types and relationship types, named connection types, named neighborhood types that are frequently used, and neighborhood schemas. An information structure management system must allow for such definitions.

An information structure management system with these capabilities will put a powerful tool in the hands of many different kinds of users.

NOTES

Bates, Marcia J. [1989]. "The Design of Browsing and Berrypicking Techniques for the Online Search Interface." *Online Review*, 13(5): 407-424.

Carlson, David A. & Sudah, Ram [1990]. "HyperIntelligence: The Next Frontier." *Communications of the ACM*, 33(3): 311-321.

Croft, Bruce W. & Turtle, Howard [1989]. "A Retrieval Model for Incorporating Hypertext Links." *Hypertext '89 Proceedings*, New York: Association for Computing Machinery; 1989.11: 213-224.

Frisse, Mark E. [1988]. "Searching for Information in a Hypertext Medical Handbook." *Communications of the ACM*, 31(7): 880-886.

Frisse, Mark E. & Cousins, Steve, B. [1989]. "Information Retrieval from Hypertext: Update on the 'Dynamic Medical Handbook' Project." *Hypertext '89 Proceedings*, New York: Association for Computing Machinery; 1989.11: 199-212.

Frisse, Mark E. & Cousins, Steve B. [1992]. "Models for Hypertext." *Journal of the American Society for Information Science*, 43(2): 183-191.

Halasz, Frank G. [1988]. "Reflections on Notecards: Seven Issues for the Next Generation of Hypermedia Systems." *Communications of the ACM*, 31(7): 836-852.

Marchionini, Gary & Liebscher, Peter [1991]. "The 'Jump' and 'Bring' Metaphors for Interaction with Hypermedia Systems." Work in progress.

Thompson, R.H. & Croft, W.B. [1989]. "Support for Browsing an Intelligent Text Retrieval System." *International Journal for Man-Machine Studies*, 1989; 30: 639-668.

Tompa, Frank Wm. [1989]. "A Data Model for Flexible Hypertext Database Systems." *ACM Transactions on Information Systems*, 7(1): 85-100.

COMPUTER SUPPORT TOOLS FOR INDEXERS

INTRODUCTION

Philip J. Smith

Various bibliographic databases exist that assist information seekers in exploring literature. These databases, containing information such as author names, book and journal titles, and abstracts of published documents, can be searched online by creating queries composed of character strings combined with various logical and positional operators.

To improve the recall and precision of such searches and to assist users in exploring related topics, most such databases include entries prepared by human indexers. These entries generally include controlled vocabulary and free-text terms assigned by an indexer or document analyst. They many also include modifications to the author's original title and a new abstract. Furthermore, the controlled vocabularies for many of these databases have been organized and integrated into thesauri.

The clear assumption behind the chapters in this section is that, for the foreseeable future, the intellectually demanding and expensive indexing process will continue to have significant value. In other words, based on both research efforts and practical experience in preparing such databases, the authors believe that human indexing will not soon be replaced by automated procedures.

Consequently, the proposals and production systems described in this section all focus on using computer support tools to enhance the performance of human indexers. This strategy for computer use is significant in and of itself, particularly given the nature and experience of the organizations supporting the work. Equally interesting is the variety of approaches taken

First, **Susanne Humphrey** describes the use of knowledge-based systems techniques to support indexing. Knowledge about terms used for indexing MEDLINE is represented in a frame system and accessed by a human in indexing a particular document. Such knowledge access is intended to increase the consistency and thoroughness of the indexer's work, and it also provides embedded training about indexing policies and practices. Equally interesting, when large numbers of documents in a database are indexed as frames, powerful "intelligent" search functions can be incorporated to assist information seekers.

John Bailey then describes a computer support system in actual use for indexing *Petroleum Abstracts*. This system, which is estimated to have increased productivity by 20 to 25 percent, uses a dictionary and thesaurus to assist indexers in identifying appropriate terminology. Sample interactions with this tool are illustrated.

Next, first **Ron Buchan** and then **June Silvester** and **Michael Genuardi** describe indexing for the NASA Scientific and Technical Information Program. Buchan's chapter focuses on the types of knowledge necessary for computer-supported indexing and on a particularly challenging application—retrospective indexing. Then, Silvester and Genuardi provide important insights into methods for developing a system for machine-aided indexing.

Finally, the chapter by **Phil Smith, Lorraine Normore, Rebecca Denning,** and **Wayne Johnson** looks at some completely different approaches to computer support for indexers. Based on empirical studies of the performances of document analysts and editors, the authors describe several classes of support tools. These include an online information access tool for the indexer (to provide access to information on indexing practices and policies), a retrieval-based training tool (in which the existing bibliographic database serves as an implicit knowledge base to provide feedback and tutoring), and a frame-based indexing tool (with strong ties to the work described earlier by Humphrey).

Chapter 8

KNOWLEDGE-BASED
SYSTEMS FOR INDEXING

Susanne M. Humphrey

INTRODUCTION

Human indexing for information retrieval is intellectually labor intensive. It requires maintaining a system of indexing rules and policies, which in turn require maintaining a controlled indexing vocabulary. These activities are being performed at the National Library of Medicine (NLM) in support of indexing the MEDLINE database using the MeSH (Medical Subject Headings) thesaurus. An additional requirement of the conventional indexing operation is maintaining and developing a user interface, known as the Automated Indexing and Management System (AIMS). As part of its intramural research program, NLM's Lister Hill National Center for Biomedical Communication is developing and testing interactive knowledge-based systems for further automating the indexing activity.

This chapter ultimately describes and recommends knowledge-based indexing, based on a unique prototype, called *MedIndEx* (Medical Indexing Expert), developed in the Lister Hill Center. Initially, we establish that in the current setting, concept indexing is needed and cannot be fully automated. Compatibility between conventional and knowledge-based indexing is then highlighted, including indexing as a cognitive process. The section on knowledge-based indexing systems describes how NLM's MedIndEx prototype addresses problems in conventional indexing. Extension of the indexing prototype to an intelligent search assistant illustrates use of the same knowledge base to integrate indexing and retrieval applications. The chapter concludes with recommendations in support of knowledge-based indexing, and the need for evaluating such systems.

NEED FOR CONCEPT INDEXING BY HUMANS

Concept indexing of document collections involves the assignment of terms to documents from a thesaurus for the purpose of retrieving these documents by subject. The expectation is that documents on the same topic may be reliably retrieved by the same indexing terms, within some reasonable range, regardless of the terminology in the documents themselves.

All database producers performing concept indexing have the same basic motivation: to provide access to information in these databases via information retrieval systems. In order to make biomedical information accessible in this way, NLM has indexed since 1966, using its MeSH thesaurus, more than 6.6 million citations covering the periodical biomedical literature. In 1990, NLM indexed nearly 400,000 documents for MEDLINE. More than two million online searches were performed in the current MEDLINE database (the most recent 2 to 3 years) using NLM's retrieval system (i.e., not counting additional searches using commercial retrieval systems that lease this database). Among NLM account holders surveyed, 96 percent searched for information by subject rather than by entering authors' names or journal titles.[1] Interestingly, nearly 60 percent of respondents wanted to retrieve all relevant citations, rather than just a few, from a particular period.

A running controversy in information retrieval focuses on whether concept indexing is even worthwhile considering the expense and labor intensiveness of thesaurus-based indexing, especially given the alternative of free-text searching where text is already available and requires only automated generation of indexes. Fidel has summarized this controversy as unresolved by studies that have attempted to determine whether textwords or thesaurus terms provide better retrieval.[2] She cites several studies that support the complementary nature of textword and thesaurus-based searching, demonstrating that neither clearly outperforms the other. The combined approach of full-text for recall and high-quality indexing for precision was recommended in a recent, extensive retrieval study by McKinin et al., testing the relative efficacy of indexing terms and full-text for MEDLINE documents.[3] Lending support for concept indexing are several recent efforts, for example, in connection with software reuse libraries,[4-7] software requirements analysis in support of knowledge-based test plan generation,[8, 9] humanities applications,[10, 11] hypermedia applications,[12] and management information systems for corporate decision making.[13]

A follow-on question might be, can thesaurus-based concept indexing be automated? Several studies have shown that it cannot. In particular, in sharp contrast to the study by Montgomery and Swanson, which concluded that 86 percent of computer indexing of titles in NLM's published *Index Medicus* would be the same as assignments by human indexing,[14] O'Connor found that correlation between indexing term and title ranged from 13 to 68 percent.[15] Another report by O'Connor studied the difficulties in devising computer rules for machine identification of the topic "toxicity" in documents from a pharmaceutical retrieval system.[16] McAllister tested a multiple linear regression model and a Boolean combinatorial model for indexing documents on instrumentation in government R&D reports, concluding that indexers in general do *not* index technical text in a machine-like fashion and that neither model is useful as a general predictor of human indexing.[17]

A few computerized indexing assistants in technical domains have been developed that use either rule-based phrase-matching or statistical correlation between humanly assigned terms and terms in text in previously indexing documents.[18-22] In no case do they produce indexing that can be used without further editing by humans. Using machine duplication of expert human indexing as a test, they apparently work better, i.e., in the 75 percent range, in easy sections of the domain, but remain at around 50 percent or less in others. Statistical systems have the "training set" problem, i.e., perhaps more than half the indexing terms are not assigned often enough (this may have negative implications for use of neural nets for thesaurus-based indexing). Linguistic systems apparently do not use general, principled linguistic knowledge. Thus, the rule base has potential of becoming quite ad hoc and therefore difficult to manage and inherently unsatisfactory to work with as it evolves. Although not creating new knowledge representations for their domains, these phrase-matching and statistical systems were nevertheless not easily built, having required at least three years in preproduction stages.

As for possible application for MEDLINE, systems that have been developed seem restricted to abstracts or relatively short reports, whereas most MEDLINE bibliographic databases index full documents for which machine-readable text is unavailable. Furthermore, machine indexing would not support the quality needed for *Index Medicus* (presuming, in addition, distinguishing central-concept terms), which is still in demand internationally.

CONVENTIONAL INDEXING

Given that indexing is likely to continue, the problem to be addressed is indexing quality. This has been measured traditionally by the degree to which searchers ultimately may rely on assignment of like headings to like concepts. A discussion of MEDLINE studies reported indexer consistency for assignment of headings to be under 49 percent.[23] Retrieval studies in which failure analysis was performed can provide further insights. For example, Lancaster found that indexing contributed to 37 percent of recall failures and 13 percent of precision failures.[24] A major source of indexing error was found to be omission of terms. To assist indexers in finding terms, MeSH includes common aliases to preferred MeSH terms, as well as "see-related" references between terms. The MeSH classification, published as *MeSH trees*, is also useful. But in a production setting it is impractical for indexers to consult MeSH for all terms they might enter or for all text phrases in documents they index.

MEDLINE uses a system of *coordinate indexing*. Indexers receive thorough training in the principle and rules of coordination, supported by indexing documentation, including manuals with illustrative examples and annotation in MeSH itself. Coordination is achieved by *post-coordination* (assigning two or more indexing terms to express an indexing concept) or *precoordination* (using a precoordinated heading or building a heading-subheading precoordination).

Coordination rules make use of the MeSH classification. Thus, indexing at NLM implicitly uses a faceted approach, in particular for expressing major ideas in documents. For example, problem-procedure coordination is used for indexing a document about the use of estrogen replacement therapy for postmenopausal bone loss. This is achieved by assigning an indexing term from each of two categories: "Osteoporosis, Postmenopausal" from the *Diseases* category, and "Estrogen Replacement Therapy" from the category *Analytical, Diagnostic and Therapeutic Techniques and Equipment.*

This example also illustrates the indexing tenet of *specificity*; that is, "Osteoporosis, Postmenopausal" is the correct indexing term, not "Osteoporosis." Even so, this term may be missed in a document that never uses the word "osteoporosis," but instead uses "bone loss," despite the MeSH cross-references "Bone Loss, Postmenopausal SEE Osteoporosis, Postmenopausal" and "Menopause SEE RELATED Osteoporosis, Postmenopausal."

Coordination also may be required for completeness in describing a disease. For instance, to index spinal osteoporosis, the term "Osteoporosis" should not be coordinated with "Spine" but rather, "Spinal Diseases," indicating the disease body site. As illustrated by this example, in NLM's indexing system organ-disease precoordinate terms (found in the *Diseases* category) are considered more specific than directly corresponding organ terms (from the *Anatomy* category) for this sort of coordination.

Some assistance for coordinate indexing is explicit in MeSH annotations. For example, a note attached to "Estrogen Replacement Therapy" instructs indexers to coordinate this term with a specific estrogen term, appending the subheading "therapeutic use." General rules for forming these heading-subheading precoordinations are attached to MeSH entries for each of the 80 topical subheadings and given in tables referring to categories of heading to which subheadings may be attached.

Systems designed for use of printed tools in applying indexing rules, as just briefly described, can be troublesome. Rules might not be stated explicitly, or they might not be repeated with all the terms to which they apply. Rules may be difficult to find. Indexers usually must be aware ahead of time that a rule probably exists. Rules are given as text instructions rather than executable procedures. The MeSH classification scheme per se does not support efficient encoding of, and access to, rules. Absence of domain-specific relations between concepts hampers the use of inferences to devise encodable rules. In addition, it takes considerable time and resources to train indexers. At NLM, for example, indexers need about a year in supervised training before their work can go into the database unreviewed by senior staff.

On the other hand, conventional systems inherently use potentially encodable procedural knowledge. As described earlier, current indexing rules, which express this knowledge, use existing factual knowledge, in particular the encoded MeSH classification scheme, and relations that are at least implicit throughout the system. Potentially usable relations may be found as implied relations between MeSH categories, in subheadings that might be used for linking terms to one another, and in single precoordinated terms (e.g., "organ-disease" implies a *site-of* relation).

Recent years have brought significant improvements in the indexing interface. In 1984 indexers began indexing interactively using NLM's Automated Indexing and Management System (AIMS). Some processing had been done in batch mode previously, such as substituting preferred forms of terms for official aliases, and validation heading-subheading coordinations according to rules of permissible combinations. The new system, however, offered assistance tailored to an interactive environment. For example, in help with checktags, the system automatically added the term "Female" when the indexer added "Pregnancy," and then displayed a message asking the indexer to add "Human" or "Animal." Indexers could request displays of the MeSH scope note, annotation, or permissible subheadings for an individual term. The system could be programmed to display help messages for specified terms.

A further development occurred in 1991 following the purchase of a number of personal computers for indexers. These machines provided a multiwindow mousing environment whereby indexers could simultaneously run the AIMS window and other windows running NLM's retrieval system to search MeSH and MEDLINE, including cut and paste between windows. Plans are underway to install the indexing manual, technical notes, and other indexing tools on a server, ensuring accessibility to the most current versions. All these developments help to pave the way for a truly knowledge-based system.

Not much is known about the indexing process per se, although some information science researchers have proposed cognitive process models of indexing based on evidence from areas of text linguistics and cognitive psychology.[25, 26] Lancaster stated that no real theories of indexing exist, and that only two fundamental rules of indexing can be identified, namely conceptual analysis and translation (which includes the specificity principle),[27] also described by Hutchins in terms of linguistic processes.[28] Reports of how humans work have relied, by and large, on what indexers say they do, and on guidelines which are essentially warnings and cautions.[17, 29] These studies usually do not include much in the way of real data.

Three reports have studied indexers directly (only Crain studied NLM indexers). Test subjects in the study by Clarke and Bennett preferred not working with a system-generated list of candidate indexing terms.[30] They furthermore suggested thesaural features for improving the directory of 20,000 possible terms, which were derived from titles and abstracts. Gotoh's think-aloud protocol analysis study concluded that for the translation stage of indexing more complicated processing is carried out than those described in indexing manuals.[31] A protocol analysis pilot study conducted by Crain revealed three reasons for indexer inconsistency: unevenness in prior experience in mapping a concept to a MeSH term, the idiosyncratic rule that an unfamiliar concept should be indexed, and ambiguity in indexing tools.[32] Three heuristics used by indexers, not reflected in indexing documentation, emerged from this pilot study: decision making and generation of hypotheses about the subject matter of an article based on the title alone, generation of hypotheses about the subject of an article based on cue concepts in the text, and selective reading of an article for specific terms.

In summary, knowledge inherent in conventional systems, coupled with installation of workstation-like environments for indexers, provides an important foundation for developing knowledge-based indexing systems. While knowledge-based systems cannot remove from the indexing process the factor of human judgment, the consistent, system-initiated assistance they provide is designed to put indexers on an equal footing in terms of awareness of indexing rules. These systems therefore hasten the evening out of level of experience differences among indexers in applying rules. Results of directly studying how indexers work seem compatible with a knowledge-based model using filling of frames for indexing concepts. Conversely, studying indexers using knowledge-based systems would assist in developing models of the indexing process.

KNOWLEDGE-BASED INDEXING

We turn now to knowledge-based indexing and how it would address the aforementioned indexing problems. In general, *knowledge-based* expert systems are distinguished by their use of encoded domain knowledge, including domain-specific relationships between concepts, as well as computer-executable rules that, using this knowledge, participate both actively and specifically to the current situation in the steps of a complex intellectual task and thereby substantively advance the task to the next step. A prototype of such a system, MedIndEx, has been developed in NLM's Lister Hill Center as part of a research program to develop and test interactive knowledge-based systems.[33] The main objective of MedIndEx is to facilitate indexing that goes in the MEDLINE product. Another focus of this research has been developing intelligent retrieval systems using the same representations and environment of the indexing system (described in the following section "Knowledge-Based Retrieval Application").

In contrast to conventional indexing systems, MedIndEx uses an object-oriented approach of *knowledge-base frames*, encoding MeSH concepts and associated indexing rules as executable procedures, using explicit domain-specific *slots* (relations) linking the current frame to other frames, and linking all concepts into an *inheritance classification*. During indexing, *indexing frames* are created as instances of knowledge-base (KB) frames named by the concatenation of this KB frame-term with a unique identifier for the document it indexes. Being an instance of a KB frame means that the indexing frame inherits domain-specific slots, which contain procedures and data, from this KB frame. These slots serve as prompts for filling indexing frames, i.e., for relating a filler-term to the frame-term according to the slot-relation. Knowledge-based assistance, such as that needed to fill slots in indexing frames, is encoded in KB frames in subdivisions of slots. (As an aside, proposals for relational analysis for indexing were published as early as the 1950s, predating the introduction of semantic network models that subsequently became associated with the field of artificial intelligence.[34])

The system outputs a database comprised of indexing frames that, unlike conventional indexing, explicitly link concepts discussed in documents they index. These linkages, representing relationships already indexed, are a source of situation-specific assistance to indexers as they continue indexing the document. Indexing frames also provide information for the system to generate conventional indexing automatically. This second output would be used to evaluate the prototype and be in a form directly usable for MEDLINE retrieval were the system completed and adopted.

Following the discussion from the previous section of coordinate indexing that could use knowledge-based assistance, a procedure-type indexing frame uses a slot called "PROBLEM" as a prompt to focus the indexer on achieving the problem-procedure coordination by adding a disease term as a filler. Figure 1 shows an "Estrogen

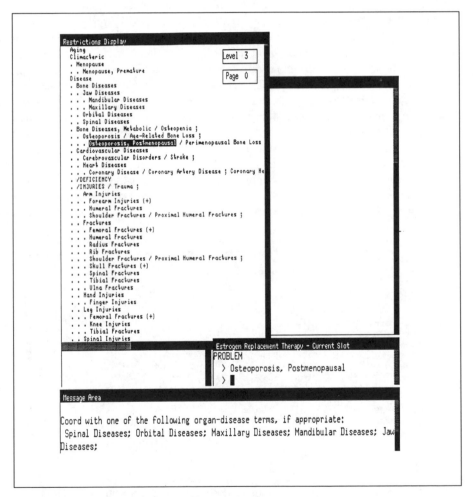

Figure 1. Indexing frame showing features of MedIndEx.

Replacement Therapy" indexing frame prompting for PROBLEM in the Current Slot window. Also shown is the display of system-initiated messages in the Message Area window; to achieve complete indexing of *spinal* osteoporosis, when an Osteoporosis term is added, a reminder is displayed to coordinate this term with one of a specific list of organ-disease terms (Spinal Diseases, etc.).

The algorithm generating this list would use several facts and inferences from the KB. This rule is encoded in a top-level KB frame, inherited by lower frames in the inheritance classification and further inherited by their instances (i.e., indexing frames). Default slot values from the KB may be displayed automatically as *candidate* fillers along with the slot prompt. Candidates may also come from previously entered fillers in indexing frames for the current document. Another form of assistance is a hierarchical display of possible fillers in a searchable, browsable, and mousable menu, as seen in the Restrictions Display window (Figure 1).

As part of system-generated conventional indexing (which can be displayed on request but over which indexers have no direct control), all heading-subheading precoordinations are formed automatically, based on completed indexing frames. Figure 2 illustrates system generation of conventional indexing, showing the user-requested Indices Display at some point during document indexing. Asterisks (stars) in this display are central concept indicators that the indexer appends. Starring and unstarring can be performed as needed during indexing, and are subject to rules that prohibit or warn against starring certain terms. The system can generate reports comparing versions of MeSH-based indexing for the same document, including computing consistency scores, as would be convenient in an experimental situation.

In summary, it is assumed that knowledge-based indexing offers indexers many advantages over conventional methods: rules as system-initiated executable proce-

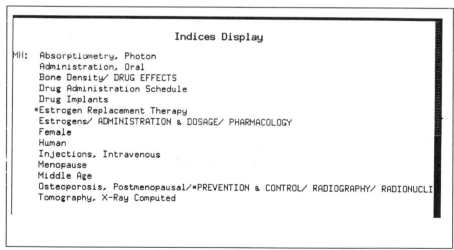

Figure 2. MedIndEx display showing system-generated conventional MeSH indexing.

dures, inheritance of this procedural knowledge for less redundancy in the KB (and therefore room for more rules) and better consistency in applying this knowledge; use of document-specific information entered previously in order to provide situation-specific assistance; and use of relationships in the domain to devise rules.

Whenever manual systems are computerized, knowledge-based systems will serve to identify gaps, inconsistencies, and vagueness of the systems they replace—an obvious benefit in any case. In addition, they would facilitate organizing the rule base leading to systematic computer-assisted development of the indexing scheme in the future. For example, when contemplating adding new indexing concepts, a knowledge-based system can be used for trying out the new term. When a term is placed in the classification scheme, the system can then show the sorts of associations and help the system would provide when the term is used in indexing. In this environment, such an action is likely to bring to the fore possible adjustments to the classification itself that might be needed. In contrast, with conventional systems it is quite problematic to evaluate the ramifications of new terms.

Managing the KB requires building a knowledge base manager tool. This would be an extension of existing thesaurus management systems, insofar as the goals of consistency and proper syntax are concerned. However, a KB, in effect, merges a thesaurus and executable indexing manual into an inheritance classification and therefore has the special requirement of checking for information that is not local to a specific structure (i.e., frame or object), but must be accessed from structures higher in the inheritance hierarchy. Managing the KB—that is, creating and editing KB frames—is more complicated than managing a thesaurus, since the knowledge engineer (formerly known as the thesaurus specialist) is responsible not only for the terms, but also encoding data and procedures needed for providing interactive indexing assistance.

General requirements of a KB are that it be consistent and have proper syntax. The KB manager we have developed checks automatically for inconsistencies and requires their resolution before the user may proceed. The system accesses inherited information, evaluates it, uses it for checking new information, and also displays inherited and local information for selection of preferred inheritance paths or for overriding inheritance with local information. Special interfaces, using menus and direct manipulation of code as objects, have been developed, not only for consistency and proper syntax, but also for ease of programming. Perhaps software can be managed by people who are not necessarily expert programmers.

The MedIndEx knowledge base contains more than 5500 frames (MeSH concepts), with 44 slots that may be filled in the indexing system, and from 1 to 12 slots for any one frame. The system, however, is designed to run similar indexing and KB manager applications in other domains. The prototype is written in Lucid Common Lisp®/Sun 4.1 (which includes CLOS) and runs on the SunSPARCstation® under the SunOS operating system. For portability, an X Window interface using X11 Release 5 and other public domain software (CLX, CLUE) has been developed. The system can be run in a client-server architecture with two SPARCstations in an Ethernet

LAN, and can be accessed from a personal computer with X server software. Remote access to the indexing system over the Internet from other parts of the United States and Canada has been demonstrated.

This environment is consistent with the conclusion of Sharif that microcomputer-based expert system shells are not suitable for developing large-scale expert systems for classification.[35] Reasons include limited size capacity and insufficient flexibility of knowledge representation structures. In preference to these tools, she recommended developing expert systems *ab initio* using languages such as Prolog or Lisp.

KNOWLEDGE-BASED RETRIEVAL APPLICATION

Paving the way toward knowledge-based retrieval, microcomputer-based systems have been developed that assist searchers offline, and then connect to online systems for search execution. Pollitt's MenUSE system is unique in displaying the MeSH classification in the form of menus as the exclusive entry to the database and the use of postings data extracted from MEDLINE.[36, 37] It then constructs and presents a menu of search statements, with postings, that reflect all possible permutations intersecting the selections. A system being developed at NLM is the Coach™ expert searcher, designed as an adjunct program to the Grateful Med® microcomputer front-end package to be invoked in the event of unsatisfactory retrieval.[38] Coach emulates a number of an expert human searcher's actions in diagnosing user search problems and determining which of a series of functions, selected by menu, to invoke for their solution. The primary knowledge source is NLM's UMLS® Metathesaurus®; also used are MeSH and special Coach knowledge sources.

KB frames and precise document representation by indexing frames as instantiations of KB frames can provide knowledge-based search tactics in an intelligent search assistant.[39] It has been suggested that development of an expert retrieval system would be a sensible outgrowth of MedIndEx research.[34, 40] In particular, adapting MedIndEx for retrieval would integrate different applications of the same KB into a single system. Also, the same interface design would be reused.

The MedIndEx prototype has been extended by adapting the document-indexing interface for use by searchers who would index their queries, producing a set of query frames. Following this, the system generates conventional indexing terms from these query frames, thereby indexing the query with MeSH terms. Then it assists searchers in expanding and combining these terms to produce a final strategy for searching the MEDLINE database. Figure 3 shows a screen from this interface, where the searcher is developing a strategy for a query involving estrogen replacement therapy. The completed strategy, which is rather sophisticated, is shown in Figure 4. This strategy can then be converted automatically to the syntax of a retrieval system and run to produce postings in a search application started by the system in another window. Although not as practical an approach, given that such databases do not exist, we have also developed a retrieval option that would search a database of indexing frames as a relational database, using an SQL query generated from the query frames.

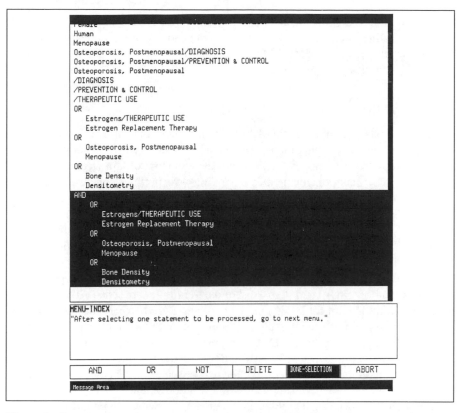

Figure 3. Developing a MeSH-based search strategy using MedIndEx for searching MEDLINE.

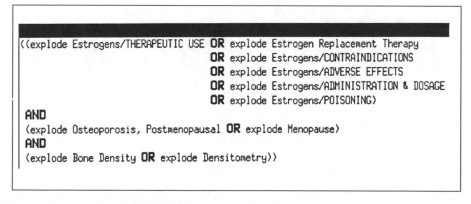

Figure 4. Completed MeSH-based search strategy from using MedIndEx for searching MEDLINE.

171

The SPECIALIST system, another knowledge-based approach being developed at NLM investigates combining sophisticated linguistic analysis with structured domain knowledge (e.g., the UMLS Metathesaurus) to improve representation and retrieval of biomedical information.[41]

CONCLUSIONS

Milstead has examined the state of the art of subject analysis (i.e., indexing) for large multidisciplinary bibliographic databases.[42, 43] Three of Milstead's recommendations—study indexing policy, look toward development of knowledge bases, and consider development of machine-aided indexing—support our conclusion to persevere in developing knowledge-based systems for machine-aided indexing.

It has been noted that the argument for frame-based indexing assistants are based on face validity, because such systems have not undergone extensive testing.[44] This point is well taken. We recognize that, although the prototype is not yet complete, an evaluation is necessary for an early assessment of the feasibility of adopting knowledge-based expert systems for indexing, and as a guide for further development.

An evaluation project to design a test of the MedIndEx approach ("Development of Experimental Design for Evaluating Knowledge-Based Expert Systems for Subject Indexing") has just been completed under contract with Herner and Company, supported by Evaluation Project 93-03, Contract N01-LM-3-3505, funded by evaluation set-aside Section 513, Public Health Service Act. In addition to Herner, other participating institutions were Decision Science Associates, Inc., and the University of Illinois Graduate School of Library and Information Science. This design is based on use of the prototype by test subjects in comparison with indexing using NLM's operational system. The methodology includes quantitative measures of indexing quality and inter-indexer consistency as well as a questionnaire. We expect to distribute this design as an NTIS report.

ACKNOWLEDGMENT

Thanks to De-Chih Chien of Management Systems Designers, Vienna, Virginia, for his work in designing, writing, and installing the interface for the MedIndEx system.

NOTES

1. Wallingford, K.T., Humphreys, B.L., Selinger, N.E. & Siegel, E.R. (1983). "Bibliographic Retrieval: A Survey of Individual Users of MEDLINE." *MD Computing*, 7(3), 166-171.

2. Fidel, R. (1991). "Searchers' Selection of Search Keys: 2. Controlled Vocabulary or Free-Text Searching." *Journal of the American Society for Information Science*, 42(7) 501-514.

3. McKinin, E.J., Sievert, M.E., Johnson, E.D. & Mitchell, J.A. (1991). "The Medline/Full-Text Research Project." *Journal of the American Society for Information Science*, 42(4), 297-307.

4.. Albrechtsen, H. (1992) "PRESS: A Thesaurus-Based Information System for Software Reuse" [Abstract]. In: N. Williamson & M. Hudon (eds.), *Classification Research for Knowledge Representation, Proceedings of the 5th International Study Conference on Classification Research* . (pp. 137-144). New York: Elsevier.

5. Durin, B. & Rames, E. (1991). "A Classification Model for Reusable Software Components." In: S.M. Humphrey & B.H. Kwasnik (eds.), *Advances in Classification Research, Proceedings of the 1st ASIS SIG/CR Classification Workshop* (pp. 47-56). Medford, NJ: Learned Information.

6.. Prieto-Díaz, R. & Freeman, P. (1987). "Classifying Software for Reusability." *IEEE Software*, 4(1), 6-16.

7. Prieto-Díaz, R. (1991). "Implementing Faceted Classification for Software Reuse." *Communications of the ACM*, 34(5), 88-97.

8. Palmer, J.D., Liang, Y. & Wang, L. (1991). "Classification As An Approach to Requirements Analysis." In: S.M. Humphrey & B.H. Kwasnik (eds.), *Advances in Classification Research, Proceedings of the 1st ASIS SIG/CR Classification Research Workshop* (pp. 129-136). Medford, NJ: Learned Information.

9. Samson, D. (1991). "Development of a Requirements Classification Scheme for Automated Support of Software Development." In: S.M. Humphrey & B.H. Kwasnik (eds.), *Advances in Classification Research, Proceedings of the 1st ASIS SIG/CR Classification Research Workshop* (pp. 147-151). Medford, NJ: Learned Information.

10. Bearman, D. & Petersen, T. (1991). "Retrieval Requirements of Faceted Thesauri in Interactive Information Systems." In: S.M. Humphrey & B.H. Kwasnik (eds.), *Advances in Classification Research, Proceedings of the 1st ASIS SIG/CR Classification Research Workshop* (pp. 9-23). Medford, NJ: Learned Information.

11. Petersen, T. (1990). "Developing a New Thesaurus for Art and Architecture." *Library Trends*, 38(4), 644-658.

12. Fidel, R. (Moderator). (1991). "SIG/CR—Indexing of Hypermedia" [Abstract]. In: *ASIS '91, Proceedings of the 54th ASIS Annual Meeting* (p. 333). Medford, NJ: Learned Information.

13. Rockmore, M. (1991). "Computer-Aided Knowledge Engineering for Corporate Information Retrieval." In: S.M. Humphrey & B.H. Kwasnik (eds.), *Advances in Classification Research, Proceedings of the 1st ASIS SIG/CR Classification Research Workshop* (pp. 137-146). Medford, NJ: Learned Information.

14.. Montgomery, C. & Swanson, D.R. (1962). "Machinelike Indexing by People." *American Documentation*, 13(4), 359-366.

15. O'Connor, J. (1964). "Correlation of Indexing Headings and Title Words in Three Medical Indexing Systems." *American Documentation*, 15(2), 96-104.

16. O'Connor, J. (1965). "Automatic Subject Recognition in Scientific Papers: An Empirical Study." *Journal of the Association for Computing Machinery* 12(4), 490-515.

17. McAllister, C.K. (1971). "A Study and Model of Machine-Like Indexing Behavior by Human Indexers." Unpublished doctoral dissertation, University of California, Berkeley.

18. Biebricher, P., Fuhr, N., Lustig, G. & Schwantner, M. (1988). "The Automatic Indexing System AIR/PHYS—From Research to Application." In: *ACM SIGIR 11th International Conference on Research and Development in Information Retrieval* (pp. 333-342). New York: Association for Computing Machinery.

19. Brenner, E.H., Lucey, J.H., Martinez, C.L. & Meleka, A. (1984). "American Petroleum Institute's Machine-Aided Indexing and Research Project." *Science and Technology Libraries*, (5)1, 49-62.

20. Buchan, R.L. (1987). "Computer-Aided Indexing at NASA." *Reference Librarian*, 18, 269-277.

21. Jacobs, C.R. (1989, January). "Machine-Aided Indexing at the Defense Technical Information Center." In: *Indexing: How It Works* (unnumbered pages). Meeting held at the National Library of Medicine, Bethesda, MD. Philadelphia: National Federation of Abstracting and Information Services.

22. Martinez, C., Lucey, J. & Linder, E. (1987). "An Expert System for Machine-Aided Indexing." *Journal of Chemical Information and Computer Sciences*, 27(4), 158-162.

23. Humphrey, S.M. (1987). "Knowledge-Based Indexing of the Medical Literature: The Indexing Aid Project." *Journal of the American Society for Information Science*, 38(3), 184-196.

24. Lancaster, F.W. (1968). *Evaluation of the MEDLARS Demand Search Service*. Bethesda, MD: National Library of Medicine (U.S.).

25. Beghtol, C. (1986). "Bibliographic Classification Theory and Text Linguistics: Aboutness Analysis, Intertextuality and the Cognitive Act of Classifying Documents." *Journal of Documentation*, 42(2), 84-113.

26. Farrow, J.F. (1991). "A Cognitive Process Model of Document Indexing." *Journal of Documentation*, 47(2), 149-166.

27. Lancaster, F.W. (1991). *Indexing and Abstracting in Theory and Practice*. Champaign, IL: University of Illinois Graduate School of Library and Information Science.

28. Hutchins, W.J. (1970). "Linguistic Processes in the Indexing and Retrieval of Documents." *Linguistics*, 61, 29-64.

29. Schwartz, C. (1977). "Indexing Behavior—Survey and State of the Art." In: *Proceedings of the 40th ASIS Annual Meeting* (part 2, fiche 8, frames D5-D14). White Plains, NY: Knowledge Industry Publications.

30. Clarke, D.C. & Bennett, J.L. (1973). "An Experimental Framework for Observing the Indexing Process." *Journal of the American Society for Information Science*, 24(1), 9-24.

31. Gotoh, T. (1983) [Jpn.] [Cognitive Structure in Human Indexing Process.] *Library and Information Science*, 21, 209-226.

32. Crain, C.J. (1985). "A Protocol Study of Indexers at the National Library of Medicine." In: J.G. Carbonell, D.A. Evans, D.S. Scott & R.H. Thomason, *Final Report on the Automated Classification and Retrieval Project: The MedSORT Project, Carnegie-Mellon University* (appendix A.4) (Contract No. N01-LM-4-3529). Pittsburgh, PA: Departments of Philosophy and Computer Science, Carnegie-Mellon University.

33. Humphrey, S.M. & Chien, D. (1990). "The MedIndEx System: Research on Interactive Knowledge-Based Indexing and Knowledge Management" (Technical Report NLM-LHC-90-03). Bethesda, MD: National Library of Medicine (U.S.). (Distributed by NTIS, Springfield, VA, Publication No. PB90-234964/AS)

34. Humphrey, S.M. (1989). "MedIndEx System: Medical Indexing Expert System." *Information Processing & Management*, 25(1), 73-88.

35. Sharif, C.A.Y. (1988). "Developing an Expert System for Classification of Books Using Micro-Based Expert System Shells" (British Library Research Paper 32). London: British Library Research and Development Department. (Distributed by British Library Publication Sales Unit, Boston Spa, Wetherby, West Yorkshire)

36. Pollitt, A.S. (1988). "A Common Query Interface Using MenUSE—A Menu-Based User Search Engine." In: *Online Information 88: Proceedings of the 12th International Online Information Meeting* (pp. 445-457)). Medford, NJ: Learned Information.

37. Pollitt, A.S. (1989). *Information Storage and Retrieval Systems: Origin, Development and Applications*. Chichester, West Sussex, England: Horwood. (Distributed by Wiley, NY)

38. Kingsland, L.C., III, Harbourt, A.M., Syed, E.J. & Schuyler, P.L. (1993). "Coach™: Applying UMLS Knowledge Sources in an Expert Searcher Environment." *Bulletin of the Medical Library Association*, 81(3), 178-183.

39. Smith, P.J., Shute, S., Galdes, D. & Chignell, M. (1989). "Knowledge-Based Search Tactics for an Intelligent Intermediary System." *ACM Transactions on Information Systems*, 7(3), 246-270.

40. Humphrey, S.M. (1989). "A Knowledge-Based Expert System for Computer-Assisted Indexing." *IEEE Expert*, 4(3), 25-38.

41. McCray, A.T., Aronson, A.R., Browne, A.C., Rindflesch, T.C., Razi, A. & Srinivasan, S. (1993). "UMLS® Knowledge for Biomedical Language Processing." *Bulletin of the Medical Library Association*, 81(3), 184-194.

42. Milstead, J.L. (1990, September). "Methodologies for Subject Analysis in Bibliographic Databases." Background paper and report of meeting sponsored by the International Atomic Energy Agency and the Energy Technology Data Exchange. Brookfield, CT: The JELUM Company. (Distributed by Department of Energy, Office of Scientific and Technical Information, Oak Ridge, TN, Publication No. ETDE/OA-58)

43. Milstead, J.L. (1992). "Methodologies for Subject Analysis in Bibliographic Databases." *Information Processing & Management*, 28(3), 407-431.

44. Smith, P.J., Denning, R., Shute, S.J. & Normore, L.F. (1991). "Toward the Development of Semantically-Based Search Systems." In: S.M. Humphrey & B.H. Kwasnik (eds.), *Advances in Classification Research, Proceedings of the 1st ASIS SIG/CR Classification Research Workshop* (pp. 153-157). Medford, NJ: Learned Information.

Chapter 9

COMPUTING SUPPORT FOR
INDEXING AT PETROLEUM ABSTRACTS:
DESIGN AND BENEFITS

John A. Bailey

INTRODUCTION

The Petroleum Abstracts Service (a Division of The University of Tulsa) markets secondary information services and products to the petroleum exploration and production industry. Since 1961, the service has published the weekly *Petroleum Abstracts* bulletin. In 1965, the service established the retrieval service that today supports the TULSA file on ORBIT and the PEP file on Dialog.

The retrieval service is based on indexing selected from a controlled dictionary of index terms. From the beginning, the service has relied on computing support that changes to reflect the progress of computing technology. Today, the indexing process is supported by an on-line system that provides various aids to the indexer. This chapter describes that system. The dictionary plays a key role in the indexing process and is the foundation for the indexing support system. Therefore, we describe the dictionary in some detail. Then, we briefly trace the evolution of computing support for indexing at the Petroleum Abstracts Service.

The main part of the chapter describes the current version of the support system (called Pedernales and abbreviated PDL). The chapter closes by discussing the benefits the current system provides, some perceived shortcomings, and some things we hope to achieve with future development.

THE DICTIONARY

The Petroleum Abstracts Dictionary comprises a collection of tightly controlled index terms that forms the basis of our various paper-based search aids products and supports the indexing process. The current Dictionary is implemented as a computer-based management system that supports adding and deleting the index terms and their relations. Other systems serve to produce the search aids.

Each Dictionary term is assigned to one of seven vocabularies, as follows:

1. *E&P*—terms in this vocabulary appear in the periodically published *Exploration and Production Thesaurus*

2. *Geographic*—these terms make up the *Geographic Thesaurus*

3. *Chemical Names*

4. *Company Names*

5. *Geographic Supplement*—some of these terms will be moved to the Geographic vocabulary before the next publication of the *Geographic Thesaurus*

6. *New E&P Terms*—these terms represent the dynamic portion of the *Exploration and Production Thesaurus* and will be moved to the E&P vocabulary before the next publication of the thesaurus.

7. *Oil Field Names.*

Each term in the Dictionary has several attributes associated with it, the most important being the validity attribute. A *valid* term is a normal term that, usually, can be used for indexing. An *invalid* term is a recognized synonym that must have a pointer to the preferred, valid term. We informally refer to an invalid term as a use term, after this required relationship.

In addition, each term in the dictionary can have several relationships assigned to it, as follows:

Broad Term—points to the parent term of the given term in a hierarchical structure

Narrow Term—a list of child terms to the given term in the hierarchy

See Also Terms—a list of additional terms that the indexer or searcher may wish to consider (Often these terms are in a similar technical area but are in different hierarchies.)

Use Term—a required relationship for an invalid term that identifies the term substitute (Occasionally, a second, valid term must be used to index the concept. In that case, that term is identified as a plus term.)

Used For—reverse pointers from a valid term to the invalid terms it represents

Scope Notes—text that defines, limits the use of, or expands on the meaning of the given term (There are several types of scope notes. The most important of these are Type 1, which provides the full spelling of terms that have been abbreviated because of a 26-character length limitation; and Type 2, which provides directions to the indexer or the searcher regarding specific uses of the term.)

Figure 1 illustrates a term from the dictionary and its several relationships. The scope note shown is a typical Type 2 along with a historical note.

A BRIEF HISTORY

Indexing at the Petroleum Abstracts Service began in 1965. From the beginning, computer support was integral to the process. The following paragraphs provide the historical perspective for the development that is central to this chapter.

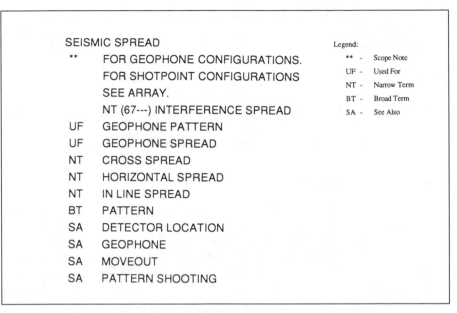

Figure 1. Example Index Term with Relationships

1965-1976

During this period, the Dictionary resided on magnetic tape. The indexers recorded their assigned terms on keypunch forms. Periodically the terms were transferred to punch cards. A monthly processing cycle consisted of the following steps:

1. Any needed modifications to the Dictionary were made using a sequential data processing technique.

2. The set of descriptor (index term) cards were matched, sequentially, against the Dictionary tape. Any descriptor that did not match, was printed out on a *bad descriptor list*.

3. Indexers would "work" the bad descriptor list, developing corrections to be keypunched and placed in the index term deck.

4. Additional terms were identified that were needed to be added to the Dictionary.

That cycle continued until the content of the bad descriptor list was reduced to zero.

1976-1981

In 1976, we implemented a disk-resident, on-line version of the Dictionary. This made it much easier to accomplish Dictionary updates. In all other respects, the processing cycle remained the same.

1981-1986

In 1981, we replaced the keypunch operation with an on-line process that provided instant verification of entered terms. This eliminated items on the bad descriptor list caused by keyboarding errors since the operator could check the term entry. This allowed us to go to a weekly processing cycle, but in all other respects the cycle remained the same.

1986-Current

The process just described was cumbersome, error prone, and entailed much duplicate effort. In this context we began, in early 1986, the development of a new support system designed to eliminate the shortcomings in the existing process. This system was called "Pedernales" and the following describes its design goals and operation.

DESIGN GOALS FOR THE NEW SYSTEM

Our objective in the design of Pedernales was to increase the productivity of our indexing process and the quality of the final result. In our effort to identify detailed design goals, a search of the literature proved fruitless. We then turned to a study done several years earlier by Martinez, et al.[1] Combining the ideas from that work with some of our own requirements, we developed the following design goals for the new system:

• As the most elementary goal, provide a way for the indexer to enter terms directly into a file and thus eliminate separate keyboarding (This would allow us to reassign keyboarding personnel while providing the indexers a better environment for recording index terms.)

• System generated instant verification of entered index terms as both a quality control process and a learning resource for the indexer. Further, if the indexer enters an invalid term (known synonym in the dictionary) he or she should be alerted to the appropriate use term(s); if the indexer enters a term that is not contained in the dictionary, he or she should be presented with a list of terms alphabetically neighboring the entered text and allowed to select the correct term with a minimum of keystrokes

• Use the on-line dictionary as a resource, replacing two printed thesauri and several other lists of index terms (Indexers should be able to select terms from a combined list—which is what the dictionary really is.)

• Provide a way for the indexer to view displays of terms related to a given term and, further, to be able to select terms from those displays with a minimum of keystrokes

• Finally, integrate Pedernales with our bulletin production process so that data such as titles and source references would have to be entered only once.

These goals became the major functional features of the resulting system, described in the following section.

SYSTEM OPERATION

Pedernales operated on a minicomputer running the UNIX operating system. In this environment, the computing system acts as a file server with the operators entering index terms into a shared file. With the current implementation, indexers use character-based terminals. Windowing is simulated by the standard UNIX screen driver, *curses*. During an indexing session, the 24-line by 80-column terminal screen is managed as two, 24-line logical windows. The left 40-column window is called the *entry* window. The right window is called the *display* window.

The system is modal; that is, it is always in one of two modes—*append* or *select*. In the append mode, the cursor is on a particular line in the entry window with the last several index terms entered displayed above it. When the indexer enters a term, the system validates the term against the dictionary and takes one of the following actions:

• Adds a valid term to the bottom of the displayed list of entered terms and the entry line is cleared, indicating that the system is ready to accept the next term

• If the term is a recognized invalid term, displays appropriate preferred valid terms (Any of those valid terms may be selected by pressing a function key.)

• If the term is not found in the dictionary, displays a list of terms alphabetically neighboring the entered text (The indexer can then position the cursor onto any of those terms and select them using a function key.)

Each term will have associated with it a term *weight*, which indicates the relative importance of the concept represented by the term to the article being indexed. Weights can range from 1 (most important) to 4 (least important). The weight defaults to 4 unless the indexer enters a lower value.

Figure 2 shows an example screen configuration in the append mode. The last five terms entered and their weights are shown at the top of the entry window, and the term being entered is seen a few lines lower. In this figure, as in all that follow, the small box symbol indicates the location of the terminal cursor.

At any point during the indexing operation, the indexer can switch to the select mode by pressing a function key. In this mode the indexer can page through the list of previously entered index terms and modify that list of terms by deletion or by changing weights. Figure 3 illustrates a typical select mode screen.

The select mode's primary use, however, is to explore the dictionary entries for any previously entered term. Using a combination of function keys, the indexer can display, in the display window, the following:

1. The "see also" terms for that term (Figure 4)

2. That term's hierarchy (Figure 5)

3. The terms alphabetically surrounding that term similar to the display provided in append mode for terms that are not found in the dictionary (Figure 6)

4. Scope notes relevant to that term (Figure 7).

In cases involving "see also" and hierarchy displays, the indexer can select a term to be added to the indexing for the current item by positioning the cursor on the term and pressing a function key.

The select mode also supports housekeeping tasks such as causing the record to be written to file and terminating work on the current item.

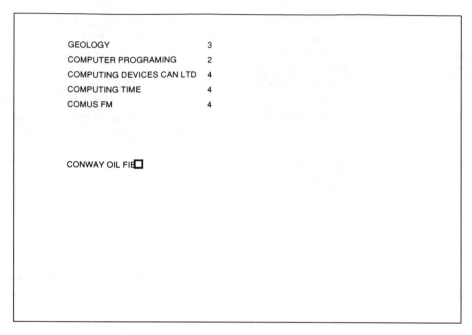

Figure 2. Append Mode Screen

DRILLING WELL CONTROL INC	4
DRILLMAC	4
DRILLOGGER	4
DRILLOSCOPE	4
DRILLING (WELL)	4
GEOLOGY	3
◙ OMPUTER PROGRAMING	2
COMPUTING DEVICES CAN LTD	4
COMPUTING TIME	2
COMUS FM	3
CONWAY OIL FIELD	3
CONCAVE	4
LENS (OPTICS)	4
TEXAS	4
OKLAHOMA	4
CALIFORNIA	4
MISSOURI	4
KANSAS	4
DRILLING EQUIPMENT	4

Figure 3. Select Mode Screen

DRILLING WELL CONTROL INC	4	CLOSED LOOP	
DRILLMAC	4	CODING	
DRILLOGGER	4	▣OMPUTER	
DRILLOSCOPE	4	LINEAR PROGRAMING	
DRILLING (WELL)	4	NONLINEAR PROGRAMING	
GEOLOGY	3		
COMPUTER PROGRAMING	2		
COMPUTING DEVICES CAN LTD	4		
COMPUTING TIME	2		
COMUS FM	3		
CONWAY OIL FIELD	3		
CONCAVE	4		
LENS (OPTICS)	4		
TEXAS	4		
OKLAHOMA	4		
CALIFORNIA	4		
MISSOURI	4		
KANSAS	4		
DRILLING EQUIPMENT	4		

Figure 4. Select Mode Screen (See-Also Display)

DRILLING WELL CONTROL INC	4	PROGRAMING	
DRILLMAC	4	COMPUTER PROGRAMING	
DRILLOGGER	4	ARTIFICIAL INTELLIGENCE	
DRILLOSCOPE	4	▣YNAMIC PROGRAMING	
DRILLING (WELL)	4	SOFTWARE	
GEOLOGY	3		
COMPUTER PROGRAMING	2		
COMPUTING DEVICES CAN LTD	4		
COMPUTING TIME	2		
COMUS FM	3		
CONWAY OIL FIELD	3		
CONCAVE	4		
LENS (OPTICS)	4		
TEXAS	4		
OKLAHOMA	4		
CALIFORNIA	4		
MISSOURI	4		
KANSAS	4		
DRILLING EQUIPMENT	4		

Figure 5. Select Mode Screen (Hierarchy Display)

DRILLING WELL CONTROL INC	4	COMPUTER ASSISTED DESIGN	
DRILLMAC	4	COMPUTER CONTROL	
DRILLOGGER	4	COMPUTER GRAPHICS	
DRILLOSCOPE	4	COMPUTER MODELING GROUP	
DRILLING (WELL)	4	COMPUTER NETWORK	
GEOLOGY	3	COMPUTER ORIENTED GEOL SOC	
C OMPUTER PROGRAMING	2	COMPUTER PROGRAM LANGUAGE	
COMPUTING DEVICES CAN LTD	4	COMPUTER PROGRAMING	
COMPUTING TIME	2	▉OMPUTER STORAGE	
COMUS FM	3	COMPUTER TERMINAL	
CONWAY OIL FIELD	3	COMPUTERIZED MAPPING	
CONCAVE	4	COMPUTING	
LENS (OPTICS)	4	COMPUTING DEVICES CAN LTD	
TEXAS	4	COMPUTING TIME	
OKLAHOMA	4	COMSTOCK TUFF	
CALIFORNIA	4	COMUS FM	
MISSOURI	4	COMYR FM	
KANSAS	4	CON FORCE LTD	
DRILLING EQUIPMENT	4	CONANT CREEK ANTICLINE	

Figure 6. Select Mode Screen (Neighbor Display)

DRILLING WELL CONTROL INC	4	USED (65-66) COMPUTER	
DRILLMAC	4	PROGRAMMING	
DRILLOGGER	4		
DRILLOSCOPE	4		
DRILLING (WELL)	4		
GEOLOGY	3		
▉OMPUTER PROGRAMING	2		
COMPUTING DEVICES CAN LTD	4		
COMPUTING TIME	2		
COMUS FM	3		
CONWAY OIL FIELD	3		
CONCAVE	4		
LENS (OPTICS)	4		
TEXAS	4		
OKLAHOMA	4		
CALIFORNIA	4		
MISSOURI	4		
KANSAS	4		
DRILLING EQUIPMENT	4		

Figure 7. Select Mode Display (Scope Note Display)

BENEFITS AND SHORTCOMINGS

We would have preferred to present objective data describing the results of this development. Unfortunately, we cannot due to a lack of baseline data and the difficulty of measuring the real benefits of this on-line support system. Instead, we offer what we hope is the next best thing—opinion of the indexers who use the system. Their evaluation of the system, both positive and negative, as well as several ideas for future development, follow.

In the positive column, the following represent the advantages of the PDL system:

• Productivity has improved. (This derives from both an increase in the rate of indexing as well as the elimination of the follow-on keyboarding step. Other difficulties, such as the legibility of the terms recorded on the indexing sheets, have been eliminated. Also, the immediate verification of entered terms has eliminated the iterative nature of earlier designs. A 20% to 25% improvement in productivity was one user's estimate.)

• Quality has improved. (The system makes it easier to check the indexing for both the indexer and the professional who provides a quality assurance check of the item. The iterative nature of earlier systems was believed to be a source of errors.)

• The quality of the work environment has improved. (Presenting the dictionary to the indexer as a single list of terms instead of several printed documents has stopped the shuffling of paper. And the system's capability to support telecommuting makes it possible for some full-time indexers to work at home a day or two each week and our contract indexers to do all their work at home.)

Nothing is perfect, of course. The system users cite the following disadvantages:

• Some discomfort with the windowing nature of the user interface. (One of the problems is the terminal screen, which simply cannot present as much information as can an open book at any given time. In particular, indexers must request for display important scope notes that are evident in an open book.)

• Occasional system failures that have required the re-indexing of some items.

• Some concern with the potential for electronic monitoring of performance and some perceived pressure to increase work rates because of the system.

• Without a paper trail, more difficulty for the person checking the indexing to inform the indexer of errors.

• On the other hand, some concern exists that too much paper still flows through the process.

FUTURE DEVELOPMENT

The several years' experience with the use of PDL and recent technical developments in computing produced more ideas for system improvement, which we have begun developing in 1992.

• Reimplement the system using a graphical user interface (GUI)
• Automatically present important scope notes with the entry of each term

- Provide key word out of context access to index terms
- Present an image of the relationships for a given term instead of separately requested relations for that term.

The GUI approach could help alleviate problems with the amount of information the indexer can see on the screen as well as improve the comfort of the working environment. Given our UNIX environment, we have chosen the X-Windows System developed at MIT (X11) as our GUI.

While the automatic presentation of scope notes is technically possible with the current system, the increased bandwidth provided by the network-based X11 is required for its practical application. Unfortunately, the indexers who work at home will not be able to use this technology due to the limited data bandwidth deliverable to most residences. They will continue using the current version of Pedernales until X11 services can be delivered over residential telephone lines. This could be economically feasible within the next three to four years.

Making keyword out of content accessibility is intended primarily for new indexers as an aid in learning about our rules for term inversion. For example, it could help the indexer determine that "MEXICO GULF" is our term for the common concept "GULF OF MEXICO."

Finally, the mouse-driven pointer capability of X11 would make it much easier to select related terms for inclusion in the indexing record. This would require the system to support the display and modification of a given record on multiple screens.

NOTES

1. Martinez, S.J. and D.W. Wattenbarger. "Requirements for a New Indexing System." *Petroleum Abstracts* Unpublished Working Notes, Tulsa, OK, ca. 1976.

Chapter 10

COMPUTERIZED DEVELOPMENT
AND USE OF THE NASA THESAURUS

Ronald L. Buchan

INTRODUCTION

Computers play a critical role in providing indexers with access to thesauri, just as they help the information seekers who search databases. They also are increasingly important in supporting retrospective indexing and monitoring quality control problems. This chapter reviews the history of the *NASA Thesaurus* and the development of computerized tools that help create and utilize the thesaurus.

FROM NACA TO NASA

The *NASA Thesaurus* lineage traces to the National Advisory Committee for Aeronautics (NACA) and its publication in 1916 of the "Nomenclature for Aeronautics (Report No. 9)" containing 9 pages and 114 terms and use references. The fifth edition, the *NASA Thesaurus* (SP-7064) consists of 3 volumes totaling nearly 1500 pages. Volume 1 contains over 21,000 terms and use references, Volume 2 contains almost 43,000 permuted entries, and Volume 3 contains over 3,200 definitions. The *NASA Thesaurus* family now includes the *NASA Thesaurus Astronomy Vocabulary* and the *NASA Thesaurus Aeronautics Vocabulary* "spinoffs" with thesauri in progress for bioastronautics and for systems engineering versions.

Other *NASA Thesaurus* computerized products include the *NASA Combined File Postings Statistics*, or CFPS, (with over 20 million postings), the two volume *NASA Thesaurus Hierarchical Update*, the cumulative *NASA Thesaurus Supplement* and its spinoff *Terms Added* (for indexers). Other thesauri are based in part on the *NASA Thesaurus* such as the *NASA News Releases and Speeches Thesaurus*, the *Thesaurus of Thesaurus Terms* (TOTT), the *Rockwell Thesaurus of Vocabulary Terms*, *McAIR Thesaurus*, and the *Thesaurus of Aerospace Terms* (TOAST). Additionally, the *NASA Thesaurus* provides subject access to *Scientific and Technical Aerospace Reports* (STAR) with over 5 million postings and to *International Aerospace Abstracts* (IAA) with nearly 10 million postings. SCAN, UPDATE, the *Continuing Bibliographies*, and the *Index to NASA Techbriefs* all use the *NASA Thesaurus* to provide subject access. The online NASA Thesaurus drives subject access and creates postings for the NASA STI Database with 20 million postings and the Aerospace Research Information Network (ARIN) with many thousand postings. The Aerospace Database, online and on CD-ROM, also uses the *NASA Thesaurus* as does the European Space Agency's Information Retrieval System (ESA-IRS). Machine Aided Indexing (MAI) and Knowledge Base Building & Maintenance Systems are intertwined using the *NASA Thesaurus* as output. The Frequency Command for subject term listings by sets presents a unique computerized approach to subject term analysis. These and other topics will be covered from the point of view of *NASA Thesaurus* computerization and use.

THE *THESAURUS OF ENGINEERING AND SCIENTIFIC TERMS*

Of the almost 300 participants, staff, and contractors working on Project Lex, Charles W. Hargrave was the only NASA participant in the *Thesaurus of Engineering and Scientific Terms* project, also known as the *TEST Thesaurus*. His participation gave a NASA background for thesaurus building. For over 20 years C.W. Hargrave also gave seasoned guidance to the *NASA Thesaurus* development. The *TEST Thesaurus* activity greatly influenced the development of the first and subsequent editions of the *NASA Thesaurus*. The leading lexicographical consultant for both thesauri was Eugene Wall of LEX Inc. Many other thesauri drew from the *TEST Thesaurus* for terms and hierarchies, building on or augmenting the core of their own nomenclatures. The *TEST Thesaurus* was issued in 1967 in two identical editions, except for binding, title page, and introduction. One was issued by the Department of Defense and the other by the Engineers Joint Council.

EARLY EDITIONS OF THE *NASA THESAURUS*

The first edition of the *NASA Thesaurus*, called the Preliminary Edition, December 1967 (SP-7030), consisted of three volumes. Although its production was computerized, its two-column format was not typeset and required two volumes for its 859-page "Alphabetical Listing." The *NASA Thesaurus* was constructed according to the principles that were to become the first American National Standards Institute thesaurus standard. It was entitled "Guidelines for Thesaurus Structure, Construction and Use" (ANSI Z39.19-1970). NASA was represented on the thesaurus standard committee by Philip F. Eckert.

The first edition lacked full hierarchies (generic structures) for all terms in Volume 1 and Volume 2; it used only Broader (BT) and Narrower (NT) relationships. The RT (Related Term) and UF (Used For) relationships were the same as used in all editions of the *NASA Thesaurus*. From the beginning, the Array Term convention was used for certain general types of terms such as "aircraft" and more specific terms were suggested for use. Today there are some 1,500 Array Terms, which are very helpful in indexing and use of the *NASA Thesaurus.*

The third volume contained four appendices. The "Hierarchical Display" (Appendix A) gave full indented hierarchies for top terms only. Appendix B provided a computerized thesaurus "Category Term Listing," which has not been provided in subsequent editions. The inclusion of thesaurus category listings has been discussed for some time and may be added in a future edition of the *NASA Thesaurus.* The "Permuted Index" (Appendix C) as well as all of Volume 3 were excluded in the abbreviated, single volume, of the second edition of the thesaurus entitled "NASA Thesaurus Alphabetical Update" (SP-7040). It was a three-column, 622-page volume that updated the first two volumes of the first edition. A listing of "Postable Terms" (Appendix D) without cross-references completed the third volume of the first edition. The third edition, referred to as the 1976 edition (SP-7050) consisted of Volume 1, "Alphabetical Listing" and Volume 2 "Access Vocabulary." This three-column, computer-typeset thesaurus of 1,262 pages established the basic format for the first two volumes of subsequent editions.

THE "BLACK" THESAURI

A now familiar black binding covers the exterior of all *NASA Thesaurus* editions beginning with the fourth edition in 1982 (SP-7051). With this edition the title of Volume 1 was changed to "Hierarchical Listing" to reflect more properly its contents. Additional terms to provide the latest terminology and systematic revisions of the hierarchies were the standard. The distinctive silver lettering became more prominent with the introduction of the thumb-indexed, bound, indexer edition. Distressed copies of the 1976 edition that were falling apart and copies with tattered hand applied letters, in lieu of thumb indexing, prompted the thumb-indexed, bound, limited edition for indexers. This edition has proven to be the best investment for indexer satisfaction, and per volume cost is minimal. It was updated with the use of a keypunch, while terminal-entered batch processing was introduced between the fourth and fifth editions.

A prototype for a typeset supplement to the *NASA Thesaurus* was presented to NASA management in July of 1983 and the first supplement was introduced in January 1984. It consisted of three parts: "Hierarchical Listing", "Access Vocabulary," and "Deletions." Three Supplements were published before the introduction of the 1985 edition (NASA SP-7053). The first supplement to SP-7053 introduced the long-awaited computerization of the definitions file of NASA-produced thesaurus term definitions. "Supplement Definitions" became Part 3 and the former Part 3, "Deletions," became Part 4, "Changes." Although the "Definitions" were computerized, they were not typeset initially nor integrated with the rest of the thesaurus file. The second supplement added *NASA Thesaurus* term definitions from the Department of Energy and began adding term definitions from NASA SP-7, "Dictionary of Technical Terms for Aerospace Use." Definitions from outside sources were edited for

NASA usage, which usually meant changing term definitions from singular to plural. By Supplement 4 for January 1988, the American Society for Testing and Materials (ASTM) had agreed to let NASA use its definitions that matched NASA terms. The ASTM definitions were selected from the *Compilation of ASTM Standard Definitions, sixth edition,* 1986. All *NASA Thesaurus* definitions and scope notes were added to the NASA STI Database for online access in July of 1988 by using the command D(isplay) Def/TERM. Rapid identification of defined terms is available internally on a regular basis and may someday be available on CFPS.

The first separately published volume of "Definitions" appeared in 1988 in Volume 3 of the sixth edition (SP-7064) of the *NASA Thesaurus.* This computer typeset volume of 150 pages contained boldfaced entries and computer generated headings. The first two volumes remained the same in scope and format but included new terms, additions to hierarchies, and revisions and changes where appropriate. The sixth edition was updated by five supplements. The first supplement added definitions from IEEE's *Standard Dictionary of Electrical and Electronics Terms, fourth edition,* 1988. In March 1990 the third supplement included terms from the American Geological Institute's *Glossary of Geology, third edition,* 1987. A very useful enhancement was the inclusion of computer identified and computer produced boldfacing of terms in the text for definitions that are defined elsewhere in the "Definitions" volume. This feature was further enhanced in Supplement 5 with the introduction of computer-aided editing of multiple and variant thesaurus terms in the definitions text. As new terms are added approved definitions sources are checked for additional standardized definitions.

NASA THESAURUS "SPINOFFS"

The *NASA Thesaurus Astronomy Vocabulary* (SP-7069) was developed with the suggestion and aid of NASA Librarian Adelaide Del Frate, for presentation at the International Astronomical Union Conference in Washington, DC in 1988. The astronomy content of the *NASA Thesaurus* benefited greatly from the work done on the astronomy minithesaurus. Although only 112 pages, it became a prototype for future minithesauri. "Spinoff" development was made possible with the use of *NASA Thesaurus* "Subject Categories" assigned to each term. Tailor-made groupings by subject were written into a program developed to translate such information into minithesauri. In addition, working slots were created with unused thesaurus category numbers whereby terms from adjacent categories could be manipulated by computer for exclusion or inclusion.

The *NASA Thesaurus Aeronautics Vocabulary* (TM-104230) represents a more refined product simply because of the experience gained in producing the Astronomy thesaurus. The Aeronautics thesaurus appeared in January 1991 and contains 222 pages. Work is currently underway to produce a Bioastronautics thesaurus and a Systems Engineering thesaurus. All these "spinoffs" benefit the basic *NASA Thesaurus* through the research necessary to produce more extensive and accurate coverage of terms.

THE ONLINE NASA THESAURUS & CFPS

The online and the printed *NASA Thesaurus* can be used hand in hand. Use of either form is dictated by requirements regarding currency (the online thesaurus is updated daily) or convenience, such as browsability and planning of subject search strategies.

As mentioned earlier, the Frequency Command is one of the most powerful search tools. It displays the frequencies of *NASA Thesaurus* terms for sets under 500. For bigger sets, the program arbitrarily selects 500 citations for subject analysis. This feature is only available on the NASA STI Database. There are two *NASA Thesaurus* expansions online: the alphabetical and the thesaurus structure. Both expansions give up-to-date information about total postings for every file that uses the *NASA Thesaurus*. The CFPS groups totals by STAR, IAA, COSMIC (Computer Software Management and Information Center), and OTHER for remaining files. This means that in some cases the CFPS is more useful for quick reference when searchers are interested in the postings picture of a term. The alphabetical expansion gives "Used For" terms that show no postings and that must be secondarily expanded to find the main term.

Using the printed thesaurus eliminates the roundabout nature of "Used For" term expansion. The expansion of the thesaurus structure for a term results in a complete picture of a thesaurus term including postings. The only thing lacking in the online thesaurus is the full hierarchy or generic structure. Studies are underway to provide the full hierarchy online. The aerospace database provides thesaurus information for STAR and IAA only without full hierarchy or postings to other NASA STI database files.

Term Weighting

At the point of indexing, term postings are determined to be either *major* or *minor*. These distinctions are meaningful in selective searching when the most important references are desired. This can be achieved by simply selecting only "Major Terms." Use of the Frequency Command provides another type of term weighting by showing the most frequently used terms in a set and displaying terms most likely to be encountered on the topic. Term weighting provides the best retrieval precision for the most relevant documents.

Term Selection

Term selection is a complex process requiring a lot of research, analysis, knowledge of the NASA and other thesauri, access to good suggestions from a variety of sources, and completion of a thorough literature review by the lexicographer. The "NASA Thesaurus Term Review" form has been used successfully for many years and cited as an example of good practice by the American National Standards Institute (ANSI) and the National Information Standards Organization (NISO) thesaurus standards. The review form asks for pertinent information helpful for determining and adding terms, hierarchies, scope notes, and definitions to the *NASA Thesaurus*.

RETROSPECTIVE INDEXING (RI) & QUALITY

The question of quality in a thesaurus and its database host is one of continued importance to lexicographers and indexers. The indexer relies on the thesaurus to fulfill needs. Indexer input is important. Inconsistencies in the thesaurus should be brought to the lexicographer's attention because continually improving the quality of a thesaurus is an important goal.

Adding new terms to the *NASA Thesaurus* (or any thesaurus) inevitably means that many existing records must be indexed to the new term. Retrospective Indexing was developed in 1984 to computerize the addition of new terms to old records. This is accomplished by searching the NASA STI Database and converting the results into transactions to add new terms to old records. This program has been so successful that in one month 26,000 RI transactions were made. [See Buchan, 1985 and 1990 in bibliography for further details.] The ability of RI to move and change large amounts of data has led to the extension of RI to other types of indexing "clean up." Specifically RI has been adapted to standardize NACA (National Advisory Committee for Aeronautics) report numbers. Other NASA STI Database files could also be standardized using the RI technique.

With RI, not only can new terms be added to old records, broader terms no longer appropriate can be deleted. Editing of major and minor term assignments can be done with the aid of the computer. Safeguards in the NASA Thesaurus maintenance programs prevent the addition of unauthorized terms to hierarchies. Terms must be created, assigned to categories, given in upper/lower case form, and inserted in the generic structure with all BTs and NTs that are appropriate. Hierarchy omissions will lead to error statements. Reciprocal relationships are always added both ways by the computer, improving quality and eliminating incomplete reciprocal relationships. Quality thesaurus work requires a constantly updated reference collection.

THE *NASA THESAURUS* EXCHANGES & ORGANIZATIONAL INVOLVEMENT

NASA has been exchanging information with other lexicographers for some 12 years. This activity provides a useful collection of other thesauri and related materials. More than 30 institutions now exchange thesauri with NASA. These organizations include Rockwell, McDonnell Douglas, the Aerospace Corporation, Engineering Information, American Geological Institute, China National Committee for Natural Sciences, and Infoterm (Vienna, Austria). This material is invaluable for lexicographical work and at the same time other lexicographers worldwide could benefit from *NASA Thesaurus* input.

Similarly, the current NASA lexicographer maintains the tradition of sitting on terminology committees to exchange ideas. He has served for several years on the National Information Standards Organization (NISO) thesaurus standards committee. The lexicographer is also a member of the International Standards Organization (ISO) Terminology Committee (TC-37) and the ISO Aerospace Committee (TC-20), serving as chair of the SC-8 subcommittee on aerospace terminology. As a member of the Committee on Terminology of the American Society for Testing and Materials (ASTM), he also serves as the chair of Working Group 3 on its dictionary. In these capacities he brings expertise to NISO, ISO, and ASTM and receives for the NASA STI Program lexicographical knowledge and contacts.

Machine Aided Indexing (MAI) Topics

Since MAI and Knowledge Based Building (KBB) & Maintenance at NASA are examined in Chapter 11, suffice it to say that their contribution to subject access through natural language processing is very great. Thesaurus work and MAI work at NASA continually interface and inspire and inform each other. In particular, the KBB program that lists phrases and terms by frequency is an invaluable source of terminological information.

Dissemination of Information

The *NASA Thesaurus* drives many SDI products, including "Selected Current Aerospace Notices" (SCAN) and its customized counterpart. Both products use *NASA Thesaurus* terms in the compilation of their search strategies. These current awareness tools provide up-to-date, reoccurring searches to fulfill information needs. For those who want the best in search access, the "Demand Index" can be made for citations in major files. The search output is augmented the next day with a "Subject Index," a "Personal Author Index," a "Corporate Source Index," a "Contract Number Index," the "Report/Accession Number Index," and the "Accession/Report Number Index." The "Demand Index" can become a personal reference tool for an individual's specialty. The NASA STI Program works hard to support the work of the computerized development and use of the *NASA Thesaurus*. The dissemination of information about the *NASA Thesaurus* is accomplished through papers, presentations, and participation in various organizations. The reader is advised to consult the bibliography or contact the author for more information.

SUMMARY

This chapter has covered three main points. First, because of the tremendous value of thesauri to indexers and information seekers, NASA has a long history of developing and refining its thesaurus products. Second, this development involves significant human effort both within NASA and through participation in national and international organizations such as NISO. While computers are a valuable tool in developing such thesauri and related knowledge-bases (as discussed in the next chapter), the intellectual contribution of lexicographers (with guidance and input from practicing indexers) remains critical to assure well-structured contents. Third, computers do, however, play a major role in constructing, and in providing access to, thesauri as well as enabling retrospective indexing. In this role, computers make it possible to use thesauri much more effectively, providing rapid, up-to-date access to

index terms and their relationships, and enabling adaptation of thesauri to reflect current uses of terminology.

BIBLIOGRAPHY

"Access" [Movie about the NASA Facility]. Washington, NASA, 1975(?). Includes a clip about the 1971 NASA Thesaurus.

American National Standard: Guidelines for Thesaurus Structure, Construction and Use. New York: American National Standards Institute, ANSI Z39.19-1980, 1980, 20 p. Includes the NASA Thesaurus Term Review form in Appendix B. (The forthcoming thesaurus standard also includes NASA Thesaurus examples.)

Anjaneyulu, V. "NASA Thesaurus in Information Retrieval." In: Seminar on Thesaurus in Information Systems, jointly organized by DRTC and INSDOC, Bangalore, India, December 1-5, 1975, p. c20-c24.

Bliss, Nonnie. *NASA Thesaurus*, 1988 edition [book review], *Sci-Tech News*, April 1989, 67-68.

Buchan, Ronald L. "Aerospace Bibliographic Control. In: *Aeronautics and Spaceflight Collections* (Special Collections, vol. 3, nos. 1/2 Fall 1985/Winter 1985/86) p. 195-229, A86-20225.

___. *Cataloging Heresy: Challenging the Standard Bibliographic Product.* Bella Hass Weinberg, ed. [Book review] *Information Today*, June 1992, p. 41.

___. "Computer Aided Indexing at NASA." In: "Current Trends in Information Research and Theory." (*Reference Librarian*, No. 18, Summer 1987), p. 269-277, A88-24200.

___. "Intertwining Thesauri and Dictionaries." In: *Information Services & Use*, 9(3) 171-175, A90-21917.

___. "A Librarian's View of Technology Transfer at NASA." Special Libraries Association Baltimore Chapter Bulletin. 40(4), April/May 1991, p. 1, 3. Based in part on the author's "Technology Transfer at NASA—A Librarian's View."

___. "The Making of the 1988 *NASA Thesaurus*." NASA STI Bulletin, May 1989, p. 2. Abstract of a talk given at a National Federation of Abstracting and Information Service course at the National Library of Medicine, January 24, 1989.

___. "NASA STI Facility." *Sci-Tech News*, October 1986, p. 111-113, A87-34719.

___. "NASA/RECON: It's Unique Resources." *Sci-Tech News*. August 1989, p. 86-88.

___. "NASA STI Database, Aerospace Database, and ARIN Coverage of `Space Law'." *Journal of Space Law*, 20(1), 1992, p. 81-82. Gives frequency breakdown of "Space Law" terms.

___. "NASA Thesaurus Bibliography." Washington, D.C., NASA Scientific and Technical Information Program, February 1992, 12p. Frequently updated.

___. "Nominymics." *ASTM Standardization News*, March 1990 (Terminology Update Column).

___. "Quality Indexing with Computer-Aided Lexicography." *Information Services & Use*, V. 12, 1992, p. 77-84. Also to be published in a volume on terminology by John Benjamins.

___. "Retrospective Indexing." Talk given at the meeting of the American Society for Information Science, Automatic Language Processing SIG held at the NASA STI Facility, January 18, 1985.

___. "Retrospective Indexing (RI)—A Computer-Aided Indexing Technique." In: *TKE'90: Terminology and Knowledge Engineering*, Hans Czap and Wolfgang Nedobity (eds.). Proceedings [of the] Second International Congress on Terminology and Knowledge Engineering, University of Trier, Federal Republic of Germany. Frankfurt/M.: Indeks Verlag, 1990, p. 339-344, A91-29797

___. "Technology Transfer at NASA—A Librarian's View in Technology Transfer: The Role of the Sci-Tech Librarian." New York: Haworth Press, 1991. (*Science & Technology Libraries*, 11(2), A91-29798.

___. "Variant Terminology." In: *Special Technical Publication of the American Society for Testing and Materials, No. 1166*. Proceedings on an ASTM terminology conference held in Cleveland in June 1991, Philadelphia; 1993, p. 95-105.

___. "Visuals for the Making of the 1988 NASA Thesaurus." A lecture presented on January 14, 1989 at an NFAIS education course: Indexing: How It Works. National Library of Medicine. BWI Airport, Maryland, 56 l.

___. "Advanced Thesaurus Topics." [BWI Airport, Md., NASA Scientific and Technical Information Facility, 1983], 10 p.

___. "The NASA Online Thesaurus." [BWI Airport, Md., NASA Scientific and Technical Information Facility, 1983], 6 p.

___. "The Printed NASA Thesaurus and Its Products." [BWI Airport, Md., NASA Scientific and Technical Information Facility, 1983], 13 p.

___. "Retrospective Indexing for New Terms." *NASA STI-RECON Bulletin & Tech Info News*, November/December 1984, p. 2.

___. Special Libraries Association Baltimore Chapter Meeting Minutes, March 20, 1991. In: "SLA Baltimore Chapter Bulletin," April/May 1991, p. 10-11. Describes talk entitled "Access to Aerospace Information—On-line."

___. "18 Millionth NASA Thesaurus Term Posting." *STI Bulletin*, April-May 1990, 20(2), p. 5.

Chamis, Alice Yanosko. "Online Database Search Strategies and Thesaural Relationship Models." [Doctoral dissertation] Case Western Reserve University, May 1984, 307 p.

Clingman, W. H. *Indexing the NASA Programs for Technology Methods, Development and Feasibility*. Dallas, TX: W. H. Clingman & Co., 1972, 94 p., NASA-CR-127465, N72-28936. Describes the application of the NASA Thesaurus terms to the NASA Program Approval Documents Program (PADS).

"Controlling the Output of Mechanized Retrieval Systems—An Empirical Test of the NASA Thesaurus." Pittsburgh, PA: University of Pittsburgh, 1967, 54 p., NASA-CR-105807, N69-77222.

Day, Melvin. "Information Processing in NASA's Library." *Wilson Library Bulletin*, 41(4), December 1966, p. 396-400, 438.

Del Frate, Adelaide. "NASA Networks: The Second Time Around." In: "Sci-Tech Library Networks with Organizations (*Science & Technology Libraries*, 8(2), Winter 1987-88), p. 47-61.

Dominick, W. D. "NASA/RECON, University-level Course Development Project." [Full course lesson plans and visuals.] Lafayette, LA: University of Southwestern Louisiana, 1988, 1918 p.

Dominick, W. D. & L. Roquemore. "NASA RECON: Course Development, Administration, and Evaluation, Final Report, FY 1983-1984." Lafayette, LA: University of Southwestern Louisiana, Dept. of Computer Science, 1984, 1697 p. NASA-CR-173922, N84-34318.

Genuardi, Michael T. "Knowledge-Based Machine Indexing from Natural Language Text: Knowledge Base Design, Development and Maintenance." In: *TKE'90: Terminology and Knowledge Engineering*, Hans Czap and Wolfgang Nedobity (eds.). Proceedings [of the] Second International Congress on Terminology and Knowledge Engineering, University of Trier, Federal Republic of Germany. Frankfurt/M.: Indeks Verlag, 1990, p. 345-351. (Also NASA-CR-4523.)

Gillum, Terry L. Comments on the TEST conventions. Systems Development Corporation, n.p., n.d., p. 7-13.

Graham, Peter S. "Quality in Cataloguing: Making Distinctions." *Journal of Academic Librarianship*, 16(4), p. 213-18.

Hammond, William. "Construction of the NASA Thesaurus—Computer Processing Support." Final report: Aries Corp., McLean, Va., 1968, 60 p., NASA-CR-95396, N68-28811. Appendix C contains a letter from the NASA Thesaurus contractor, Eugene Wall, who contributed greatly to the NASA Thesaurus review. Appendix 7 presents "Thesaurus Rules and Conventions," which provided the foundation for TEST and NASA Thesaurus development.

Hammond, William, et al., [Letter to the editor]. *American Documentation*, October 1968, p. 416. States that the computer work on the first NASA Thesaurus was done by Aries Corp.

Hargrave, Charles W. & Eugene Wall. "Retrieval Improvement Effected by Use of a Thesaurus." In: *Proceedings of the American Society of Information Science 33rd Annual Meeting*, Philadelphia, October 11-15, 1970. Washington: ASIS, 1970, p. 291-294.

Hines, Theodore C. [Book review of first edition of the *NASA Thesaurus*.] *American Documentation*, April 1968, p. 208-209. Provides background information on the development of the NASA Thesaurus.

"Integrated Astronomy Thesaurus." *STI Bulletin*, 20(1), January/February/March, 1990, p. 3.

International Energy Subject Thesaurus. [Oak Ridge, TN], 1990, 1061 p.

Jack, Robert F. "The NASA/Recon Search System: A File-by-File Description of a Major—But Little Known—Collection of Scientific Information." *Online*, November 1982, 6(6), p. 40-54.

___. "The NASA Thesaurus: a Bibliography, Derived from ERIC, ISA, and LISA Databases." BWI Airport, MD, NASA STI Facility Operated By RMS Associates, August 22, 1989, 5 p.

Johns, Gerald. "Dictionary Making by Conference and Committee: NACA and the American Aeronautical Language, 1916-1934." *Journal of the American Society for Information Science*, 35(2), p. 75-81. Traces the earliest forerunner of the NASA Thesaurus, NACA's Nomenclature for Aeronautics.

Jones, Paul E. et al. "Application of Statistical Association Techniques for the NASA Document Collection." Prepared by Arthur D. Little, Inc., Washington, DC, National Aeronautics and Space Administration, February 1968, 112 p., NASA-CR-1020, N68-17154.

Kavanagh, Stephen K. & Jay G. Miller. "The Aerospace Database." *Database*, April 1986, p. 61-67, A89-31211.

Kent, Allen & Martha Manheimer. *The Applicability of the NASA Thesaurus to the File of Documents Indexed Prior to its Publication*, Pittsburgh: University of Pittsburgh, 1969, 205 p. NASA-CR-105946, N69-77899.

Kent, Allen et al. *Controlling the Output of Mechanized Retrieval Systems: An Empirical Test of the NASA Thesaurus*. Pittsburgh: University of Pittsburgh, 1969, 52 p., NASA-CR-105807, N69-77222.

___. "NASA Automatic Subject Analysis Technique for Extracting Retrievable Multi-Terms (NASA TERM) System." BWI Airport, MD: Informatics Inc., 1978, 27 p., NASA-CR-157398, N78-30992.

Kim, Chai. "Theoretical Foundations of Thesaurus-Construction and Some Methodological Considerations for Thesaurus-Updating." *Journal of the American Society for Information Science*, March-April, 1973, p. 148-156. Includes information on the *NASA Thesaurus*.

Klingbiel, P. H. "Phrase Structure Rewrite Systems in Information Retrieval." *Information Processing and Management*, 21(2), p. 113-126. A85-44770.

Lancaster, F. W. *Indexing and Abstracting in Theory and Practice*. Champaign, IL: Graduate School of Library and Information Science, 1991, 342 p. See especially Chapter 6—"Quality Indexing" and Chapter 14—"Automatic Indexing, Automatic Abstracting and Related Procedures."

___. *Vocabulary Control for Information Retrieval, 2nd ed.*, Arlington, VA: Information Resources Press, 1986, 287 p.

Landau, Sidney I. *Dictionaries: the Art and Craft of Lexicography.* New York: Charles Scribners' Sons, 1984, 378 p.

MCAIR THESAURUS. McDonnell Douglas Corporation, 1991, 702 p.

Miller, Eugene. [Letter to the editor.] *American Documentation*, October 1968, p. 416. Acknowledges roles played by Documentation, Inc., Eugene Wall and Charles Hargrave in the development of the *NASA Thesaurus.*

Morenoff, Jerome, Donald L. Roth & James W. Singleton. *Space Law Information System Design.* Rockville, MD: Ocean Data Systems, Inc., 64 p. N73-17986, NASA CR-130746.

NASA Combined File Postings Statistics Based on [the] NASA Thesaurus. January 1968-July 1992, semiannual, cumulative, v.p. Includes tabular summaries of all NASA Thesaurus Postings including the COSMIC NASA software.

"NASA Publication Features ASTM Terminology Relating to Aerospace Use." *ASTM Standardization News*, May 1989, p. 18.

NASA Subject Authority List, Alternate Data Base, Postings Statistics (1962-1967), September 1975, Reprinted 1985 with new introduction, 422 p.

NASA Thesaurus (NASA SP-7064). [Book review]. *Journal of Space Law*, June 1991, 19(1), p. 88.

NASA Thesaurus Aeronautics Vocabulary. Washington, DC: NASA, Office of Management, Scientific and Technical Information Division, 1991, 222 p. (NASA Technical Memorandum 104230), N91-16847. Contains over 4,700 terms from the *NASA Thesaurus* file in 1990 that deal with aeronautics.

"NASA Thesaurus Aids." *AIAA Bulletin.* Feb. 1990, p. B6.

"NASA Thesaurus Alphabetical Update." 1971, NASA SP-7040, 622 p.

"NASA Thesaurus Astronomy Vocabulary." Presented at the International Astronomical Union Conference, 27-31 July 1988, NASA SP-7069, 112 p., N88-24553. Contains 1600 terms and entries from the 1988 *NASA Thesaurus* that deal with astronomy.

"NASA Thesaurus Definitions Now Online." *STI Bulletin*, July 1988, 18(7), p. 3-4.

"NASA Thesaurus Hierarchical Update, A Semiannual Cumulative Listing." A complete cumulative printout of hierarchical information published in March and September, including *NASA Thesaurus* category numbers.

"NASA Thesaurus Supplement, 1984-1985, A Semiannual Cumulative Supplement to the 1985 Edition of the NASA Thesaurus." NASA SP-7053. This three-part supplement consists of 3 cumulative issues.

"NASA Thesaurus Supplement, 1986-1988, A Semiannual Cumulative Supplement to the 1985 Edition of the NASA Thesaurus." NASA SP-7053. This four- part supplement consists of 4 cumulative issues.

"NASA Thesaurus Supplement, 1989-1991, A Semiannual Cumulative Supplement to the 1988 Edition of the NASA Thesaurus." NASA SP-7064. This four-part supplement consists of 5 cumulative issues.

NASA Thesaurus: Subject Terms for Indexing Scientific and Technical Information. Preliminary Edition, 1967. NASA SP-7030. Three volumes: V. 1, Alphabetical listing, A-L; V. 2, Alphabetical listing, M-Z; V. 3, Appendices. NASA's first Thesaurus, includes listings by *NASA Thesaurus* categories in V. 3.

NASA Thesaurus. 1976 Edition, NASA SP-7050. Two volumes: V. 1, Alphabetical Listing; V. 2,

Access Vocabulary. First NASA Thesaurus to contain an Access Vocabulary and to be typeset.

NASA Thesaurus. 1982 Edition, NASA SP-7051. Two volumes: V. 1, Hierarchical Listing; V. 2, Access Vocabulary. Includes extensive revisions and new material. Typesetting was redesigned and made more legible.

NASA Thesaurus. 1985 Edition, NASA SP-7053. Two volumes: V. 1, Hierarchical Listing; V. 2, Access Vocabulary.

NASA Thesaurus. 1988 Edition, NASA SP-7064. Three volumes: V. 1, Hierarchical Listing; V. 2, Access Vocabulary; V. 3, Definitions. Definitions are presented for the first time in the new third volume. The three volumes total 1440 pages.

NASA Thesaurus. 1994 Edition, forthcoming.

[NASA] "Subject Authority List." December 1967, 422 p.

"New Thesaurus Terms." *STI Bulletin.* A periodic column.

Niehoff, R. & Mack, G. "Evaluation of the Vocabulary Switching System." Final report, 1 Oct. 1979-31 Oct. 1984. Battelle Columbus Laboratories, Ohio, 205 p.

Niehoff, R. T. "Development of an Integrated Energy Vocabulary." *Journal of the American Society of Information Science,* 27(1), p. 3-17.

Nomenclature for Aeronautics. Washington, DC: National Advisory Committee for Aeronautics, (Government Printing Office), 1916, 9 p.

"Orbital Transfer Vehicle Launch Operations Study: Automated Technology Knowledge Base, Volume 4." Final report, Kennedy Space Center, FL: Boeing Aerospace. 124 p. 1986, NASA-CR-179706, N87-10883. Includes strategy for compiling bibliographies using the *NASA Thesaurus.*

Pryor, Harold E. "An Evaluation of the NASA Scientific and Technical Information System." *Special Libraries,* V. 66, November 1975, p. 515-519, A76-11821.

___. "Managing Aerospace Information." In: *Information Systems and Networks, Eleventh Annual Symposium,* March 27-29, 1974. Westport, CT: Greenwood Press, 1975, p. 84-92.

Rainey, Laura. "Experience With the New TEST Thesaurus and the New NASA Thesaurus." *Special Libraries,* January 1970, p. 26-32.

Raitt, D.T. "Recall and Precision Devices in Interactive Bibliographic Search and Retrieval Systems." *Aslib Proceedings,* V. 32, July-August 1980, p. 281-301, A80-45699. Also looks at thesauri of aerospace interest, the European Space Agency's IRS system including the *NASA Thesaurus,* the *TEST Thesaurus,* and the *INSPEC Thesaurus.*

"RTIS: Rockwell Technical Information System; Thesaurus of Subject Terms." n.p. Rockwell International, 1991, 923 p., looseleaf.

Sauter, H. E. & L. N. Lushina. "Organizational Structure and Operation of Defense/Aerospace Information Centers in the United States of America." Washington, DC: NASA, 23 p. In: *Use of Scientific and Technical Information in the NATO Countries,* AGARD-CP-337, N83-31535.

Silvester, J. P., R. Newton & P. H. Klingbiel. "An Operational System for Subject Switching Between Controlled Vocabularies: A Computational Linguistics Approach." Contractor report, 2 Nov. 1981-31 Dec. 1983. McLean, VA: Planning Research Corp., 1984, NASA-CR-3838, N85-11903. This report describes NASA's subject switching system that feeds into the *NASA Thesaurus.*

Skolnik, Herman. [Book review]. *Journal of Chemical Information and Computer Sciences,* 16(3), p. 396.

Steinacker, Ivo. "Indexing and Automatic Significance Analysis." *Journal of the American Society for Information Science,* July-August 1974, p. 237-241.

Thesaurus of Aerospace Terms. n.p. The Aerospace Corporation Library Services, 1991, 827 p. (Report number 857-62).

Thesaurus of Scientific, Technical, and Engineering Terms. Washington, DC: Science Information Research Center, Hemisphere Publishing Corp., c1988, xix, 841, 376 p. The book jacket states that the Thesaurus is based on the indexing vocabulary developed by NASA, and also utilizes the DOD *Thesaurus of Scientific and Engineering Terms.* The thesaurus is an exact reproduction of the 1985 edition of the *NASA Thesaurus* and the first part of the first supplement except for the deletion of every NASA reference. The referenced DOD title does not exist but probably refers to the fact that the first *NASA Thesaurus* drew some terminology from the jointly produced *Thesaurus of Engineering and Scientific Terms.* The book jacket also refers to the appropriate NASA subtitles for Volume 1—Hierarchical Listing, and Volume 2—Access Vocabulary.

Vasaturo, Ronald. "What's There to NOTIS." *Technicalities*, January 1984, p. 7-9.

Waite, David P. "NASA Thesaurus." *American Documentation*, 20(2) p. 297. [Letter to the Editor].

Weinberg, Bella Haas. "Issues in the Revision of the Thesaurus Construction Standard." *Bulletin of the American Society for Information Science*, December/January 1989, p. 26-27.

Wente, Van A. & Gifford A. Young. "Operating Experience with NASA/SCAN, Selected Current Aerospace Notices, Promoting Selectivity in Information Transfer to Abstract Journals, Accession Lists and Bibliographies." In: *American Society for Information Science, Annual Meeting, Columbus, Ohio, Oct. 20-24, 1968, Proceedings, Volume 5—Information Transfer.* New York: Greenwood Press, 1968, p. 217-222.

___. "Selective Information Announcement Systems for a Large Community of Users." *Journal of Chemical Documentation,* 7(3), August 1967, p. 142-147. A68-17693 Background and analysis of NASA SCAN, NASA's SDI program.

Westbrook, J.H. & W. Grattidge, eds. "A Glossary of Terms Relating to Data, Data Capture, Data Manipulation, and Databases `being'." *CODATA Bulletin*, 23(1-2), January-June 1991. (ICSU Committee on Data for Science and Technology). Includes *NASA Thesaurus* as a source.

Wilson, John. "Machine-Aided Indexing for NASA STI." *Information Services & Use*, V. 7, 1987, p. 157-161.

Chapter 11

MACHINE-AIDED INDEXING
FROM THE ANALYSIS OF
NATURAL LANGUAGE TEXT

June P. Silvester and
Michael T. Genuardi

INTRODUCTION

The primary goal of natural language understanding is to establish a machine system that can identify the conceptual content of written text and manipulate those concepts to mimic some human intellectual activity. Although the goal has not been achieved, many natural language analysis methods developed in support of natural language research have been incorporated into operational systems that function as "computer aids" rather than as fully automatic techniques.

In the area of machine-aided indexing (MAI), numerous strategies for statistical and syntactic analysis and knowledge base design have been used. The function of such systems is to provide a concept-level analysis of the textual elements of documents or document abstracts—the final output being a list of candidate index terms from an established classification scheme or thesaurus (Figure 1). The overall performance of these MAI systems depends largely upon the quality and comprehensiveness of their knowledge bases. The term "Knowledge Base (KB)" refers here to the MAI system elements that contain the information and rules required to derive controlled index terms from text phrases. These KBs may be structured in several ways, e.g., as phrase dictionaries or as conceptual networks.

This chapter describes the functional elements of text-based MAI systems and, in particular, the MAI system implemented by the National Aeronautics and Space Administration's (NASA's) Scientific and Technical Information (STI) Program staff at the NASA Center for AeroSpace Information (CASI).[1] Second, it discusses the development and maintenance of the NASA MAI knowledge base with emphasis upon statistically-based text analysis tools designed to aid the knowledge base developer. Finally, it describes the application of these tools as an aid in thesaurus construction.

```
        TITLE   =  Mission options for an electric propulsion demonstration flight test

     ABSTRACT   =  Several mission options are discussed for an electric propulsion space test
                   which provides operational and performance data for ion and arcjet
                   propulsion systems and testing of APSA arrays and a super power system.
                   The results of these top-level studies are considered preliminary.  Ion
                   propulsion system design and architecture for the purposes of performing
                   orbit raising missions for payloads in the range of 2400-2700 kg are
                   described.  Focus was placed on a design which can be characterized by
                   simplicity, reliability, and performance.  Systems of this design are
                   suitable for an electric propulsion precursor flight which would provide
                   proof of principle data necessary for more ambitious and complex
                   missions.

          MAI   =  mission planning
    SUGGESTED      electric propulsion
        TERMS      flight tests
                   ion propulsion
                   arc jet engines
                   propulsion system configurations
                   propulsion system performance
                   payloads
                   reliability
                   proving
```

Figure 1. MAI-Suggested Terms from Natural Language Input Consisting of Title and Abstract

TEXT-BASED MACHINE-AIDED INDEXING

Functional Elements

The functional or operational elements of a text-based MAI system can be generalized as the following:

- delineation of text phrases
- identification/reduction of semantic units
- semantic analysis

The first operation has as its primary task the establishment of general boundaries or parameters to help assure that the words comprising a semantic unit actually represent a statistically valid co-occurrence or a grammatically "correct" association (such as between an adjective and the noun it was intended to modify). This can be done with techniques of varying complexity from simple phrase-breaking procedures to formal syntactic parsing. Tradeoffs may be necessary to achieve the ultimate goals of the system. Simple nonsyntactic techniques have the advantage of being time efficient; full parsing, on the other hand, may provide greater accuracy.

The second operation identifies semantic units by extracting from these phrases single words and multiword strings that may express concepts within a given subject domain. This operation typically requires a means for the successive concatenation of words within the selected phrases. Some systems may include a process for word stemming (as in "photoelectrical" to "photoelectric") or phrase normalization (as in "surface of the moon" to "moon surface").

The third operation, semantic analysis, translates the final forms of the semantic units (word strings) that are identified in the second operation into appropriate indexing terms from the controlled vocabulary. A primary function of the knowledge base is to serve as the equivalency table for this translation process.

Knowledge Base Design and Function

Controlled vocabulary terms, as contained in a thesaurus or classification scheme, express the concepts of interest within an organization's particular subject domain. This collection of controlled terms represents the core of the MAI knowledge base. The knowledge base can be viewed as a conceptual network that 1) defines the relations between natural language concepts and thesaurus terms, and 2) allows the inheritance of the thesaurus-term hierarchical structure by the natural language equivalents—a feature for possible future use at NASA.

Besides containing entries that map natural language words and phrases to controlled vocabulary terms, the knowledge base contains entries that represent decisions regarding the relevancy of particular concepts. For example, within the aeronautics domain, the concept "aircraft" is much too broad to serve as a useful indexing term for most instances where the word "aircraft" appears in text. In this case, specific entries in the knowledge base would initiate a search for a multiword semantic unit (such as "A-320 aircraft," which describes the specific vehicle in question; or "aircraft stability", "aircraft construction materials," etc., which indicate the particular aeronautical aspect of interest). Other entries in the knowledge base serve to disambiguate certain words such as "matrices," which might refer to either mathematical matrices or material matrices.

Naturally, the design of any particular knowledge base depends on how the other system operations are carried out. The procedures selected for initial phrase delineation and analysis define what kinds of information should be represented in knowledge base entries and also how large an operational file needs to be (e.g., the use of word stemming or phrase normalization can reduce the number of required entries). Likewise, the strategies used for disambiguating words and for analyzing relevancy define the level of complexity required for knowledge representation and ultimately may dictate the kind of data structure used.

NASA MAI SYSTEM

Historical Overview

In 1981, the management team for the National Aeronautics and Space Administration's Scientific and Technical Information (STI) Program decided to begin work on a Machine-Aided Indexing (MAI) system. The indexing of NASA acquisitions for its STI Database is done at the NASA Center for AeroSpace Information (CASI) in an online environment.

MAI's first goal was to automate the translation of document subject terms assigned by indexers at the Defense Technical Information Center (DTIC) to equivalent *NASA Thesaurus* terms. This capability became operational in a batch mode in June 1983.[2, 3]

The second goal was to extend this "Subject Switching" capability to documents indexed by the Department of Energy (DOE). This was done during the following year.

The ultimate goal was to use MAI, based on the analysis of natural language text, in an online mode to provide NASA indexers with a list of pertinent, candidate, *NASA Thesaurus* terms that could then be edited electronically. MAI from natural language was first tested in 1984 and became operational in a batch mode in August 1986. Online MAI was initiated in the fall of 1988.

Both the Subject Switching and natural language MAI processes are based on a "phrase structure rewrite" method, the historical development of which is described by Klingbiel.[4] Very simply, a phrase structure rewrite system, or "lexical dictionary," is a table format and an access procedure that provides an efficient means for the translation to a controlled vocabulary of any single and multiword phrases that are input into the system. The phrase-dictionary-style knowledge base that provides the *NASA Thesaurus* equivalents for selected natural language words and phrases contains more than 115,000 entries and is expected to become several thousand entries larger. It was once thought that a minimum of 120,000 entries would be necessary for successful operation, but NASA was able to go operational when the KB contained fewer than 67,000 entries.

The current implementation of natural language MAI at CASI proceeds as follows. Titles and abstracts of input document records either are received in machine-readable form, scanned, or typed into the record by someone working in the online Input Processing System (IPS). IPS is mounted on an IBM 4381 mainframe and accessed by input processing staff from an IBM 3278-4 or 3180-type terminal. NASA indexers currently have two options for accessing natural language MAI—they can use MAI online in an interactive mode, or they can transfer the document records to a queue, which is processed through MAI in a batch. Batches are run four times a day. An indexer who wants to use interactive MAI presses a function key and, within approximately 6 seconds for a title plus a 150 to 250 word abstract, receives 10 to 15 thesaurus terms for consideration. The indexer reviews these candidate terms and makes additions or deletions as needed. The system currently serves five indexers but has been stress-tested with several more.

Original Design and Revisions

The application of the original text-based MAI system was limited due to its unacceptably long response time when used with interactive workstations. An additional problem was the slow development time and level of manual effort associated with KB construction. In 1987, we began to address these problems.

Tests of the original MAI system determined that the phrase delineation process accounted for much of the total processing time. The original phrase delineation process, a second generation of the process used by DTIC in its MAI system, was based on the syntactic analysis of input text. It required that 1) the syntactic class of each word be identified from a separate table, and 2) the sequence of the resulting syntactic classes be checked against a table of "grammar" rules.

The new design replaced the syntactic procedure with a simple preprocessing step, and incorporated a proximity limit for words concatenated during the semantic-unit identification process. Preprocessing consists of breaking raw text input at end-punctuation points (period, colon, semicolon, question or exclamation mark) and where occurs any of about 250 carefully selected "stopwords." (These are discussed and listed in Appendix A.) The proximity limit is a constraint imposed on the word-concatenation process (that process functioning to identify semantic units for subsequent look-up in the KB). It is an empirically established value above which the likelihood of grammatically or statistically "incorrect" word associations becomes significant. This value was based on observations of input and it also considered speed, accuracy, and frequency of occurrence tradeoffs in the final processing.

The current routine for carrying out semantic-unit identification and analysis, Access-2, uses a proximity limit of five. Higher values were found to increase the rate of error in final output (and also added slightly to the required processing time); lower values missed too many indexable concepts. The general logic of the Access-2 program is summarized in Appendix B. For example, consider the following sentences from a document abstract on "Helicopter Noise":

> Acoustic data for a 40 percent model MBB BO-105 helicopter main rotor were obtained from wind tunnel testing and scaled to equivalent actual flyover cases. It is shown that during descent the dominant noise is caused by impulsive blade-vortex interaction (BVI) noise. In level flight and mild climb BVI activity is absent; the dominant noise is caused by blade-turbulent wake interaction.

Phrase delineation results from ending textual word strings whenever a stopword (see Appendix A) or any thought-ending punctuation such as a period, colon, or semicolon is encountered. The phrase delineation process results in the following phrases from this title and abstract:

- helicopter noise
- acoustic data for a 40 percent model MBB BO-105 helicopter main rotor
- from wind tunnel testing and scaled to equivalent actual flyover cases
- descent the dominant noise
- by impulsive blade-vortex interaction (BVI) noise
- in level flight and mild climb BVI activity
- absent
- the dominant noise
- by blade-turbulent wake interaction.

To illustrate how the knowledge base entries direct concatenations, look at the second five-word segment of the third phrase above, i.e., "wind tunnel testing and scaled." The first two words, "wind tunnel," are found as a key to an entry in the knowledge base. The posting term field for this entry contains an asterisk. This means that the system must look for these two words (wind tunnel) followed by another word. The next word in this five-word segment is "testing." When "wind tunnel testing" is looked up, the knowledge base provides "wind tunnel tests" as the equivalent thesaurus term, and this is printed out as a suggested term for the indexer to review. Other prescribed combinations of words in this segment provide no thesaurus terms as they either do not occur in the knowledge base or they translate to 00.

The use of the Access-2 program (see Appendix B) provides the following concatenations from the delineated phrases above, and the equivalents of these concatenations in NASA Thesaurus terms:

- helicopter noise *use* AEROACOUSTICS, AERODYNAMIC NOISE, AND AIRCRAFT NOISE
- BO-105 helicopter *use* BO-105 HELICOPTER
- helicopter rotor *use* ROTARY WINGS
- wind tunnel testing *use* WIND TUNNEL TESTS
- descent *use* DESCENT
- climb *use* CLIMBING FLIGHT
- turbulent wake *use* TURBULENT WAKES.

Another feature of the original design that affected overall processing time was the system of "logic codes" used to direct the word concatenation process. These codes, adopted from Klingbiel's "lexical dictionary" format, were assigned to each entry in the knowledge base.[4] Using these logic codes added to the input/output requirements of the original programs. In addition to the machine-time inefficiencies of the original phrase delineation and logic-code processes, both of these procedures were found to be unnecessary for quality performance of the system.

Another revision to the system prevented the selection of semantic units that consisted of single words, or pairs, or groups of words that were embedded in input that had already been translated. This had the effect of selecting the most specific available index terms, in conformance with NASA's indexing policy.

With the revisions in place, the NASA MAI system matched and often improved upon the output of the original system and the online response time was reduced by approximately 70 to 80 percent. With the goal of an online MAI system, this time savings was an important improvement.

The NASA system is a good example of MAI process integration. In the same way in which the phrase delineation process was incorporated into identifying semantic units, the concatenation method used for finding semantic units is integrated with the semantic unit translation process, i.e., the KB look-up. The method for the identification of semantic units is carried out using concatenation logic rules that dynamically incorporate information from knowledge base entries. Thus, in many instances, the presence or absence of specific KB entries directs the concatenation process.

KNOWLEDGE BASE DESIGN AND DEVELOPMENT

General Design and Function

In a text-based MAI system such as NASA's, semantic analysis depends upon the content of the KB. Primary concerns with this functional element were the slow development time, the level of manual effort associated with KB construction and maintenance, the need to generate high quality output, and the problem arising from limiting analysis to certain syntactic phrase forms.

The KB has a relatively simple structure. It contains two fields: the *Key* field holds the natural language input word or phrase that is also the address to the record in the computer file; the *Posting Term* field contains the semantically equivalent *NASA Thesaurus* term(s). Variant forms of a word are included in the KB keys because NASA MAI experience indicated that word-stemming tends to result in ambiguous word forms and cause unwanted output. Words containing hyphens are searched once with the hyphen and, if not found, are then searched without the hyphen. Ambiguous terms in the text may be disambiguated by the choice of longer, more informative entries in the KB, or ambiguous terms may be left out of the KB or receive a null translation for that reason.

In the semantic analysis function, the semantic units are matched with the keys in the KB and the equivalent thesaurus terms are written out. If the content of the semantic unit is not of indexable importance, the KB may delete the unit, i.e., provide no translation into thesaurus terms. The semantic value of each unit depends upon the content of the KB, which, in turn, depends upon the evaluation of entries made by the analysts who created the KB and continue to add records to it.

KB Development Tools

The MAI KB's form and content depend on the design for carrying out the basic system operations. To a large extent, a system can compensate for design tradeoffs by incorporating the appropriate class or classes of entries into the KB. As previously noted, the NASA system does not include a mechanism for word-stemming; therefore, all variant forms of words, including British spellings, are included in the KB entries. Thus, a system designer must consider the tradeoffs among the size of the KB file, the system's response time, and the level of complexity selected.

The "level of complexity" means the number of special factors to be considered before the system suggests the thesaurus terms for indexer review. These factors might include words containing embedded slashes or hyphens, upper and lower case, the coexistence of broader or narrower suggested terms from a single hierarchy, a category code, and so on.

For an online system designer, quick responses have high priority. Regardless of the specific design selected for an MAI system, its overall performance largely depends upon the quality and the comprehensiveness of its knowledge base. Strict control and input from domain experts are critical during the database development process. The resources spent in careful construction of the MAI KB will pay off with high quality output and indexer acceptance.

Most knowledge base content is represented by the entries that map natural language expressions to thesaurus concepts. The identification of those expressions may seem like an infinite task, especially in a large thesaurus. KWIC (Keyword In Context) indexes of available text are limited in use precisely because they are arranged by "key words" rather than the actual target concepts. Such indexes can identify some of the natural language expressions that may need to be translated, but not what index terms are the target. A review of abstract text on a case-by-case basis was found to be highly inefficient, besides being untargeted with regard to domain concepts.

Time consumption was not the only drawback to these two methods. Both can lead to an unnecessarily large knowledge base due to the addition of expressions that are essentially unique or have a very low frequency of occurrence. After trying the above methods, NASA developed a statistically based text analysis tool that presents the domain expert with a well-filtered list of synonymous and conceptually-related phrases for each thesaurus concept. This tool was designed to satisfy three main requirements:

1. The output phrases for any given use of this tool would be targeted to one specific thesaurus concept—thus all expressions related to a particular target term could be analyzed together. By targeting, in separate operations, all the terms in a single hierarchy, or all of the terms that share a word (such as MATRICES, MATRICES (CIRCUITS), and MATRICES (MATHEMATICS)), expressions that can lead to ambiguity can be identified and analyzed together.

2. The output phrases would be restricted to those that had a high frequency of occurrence within the existing NASA STI Database—thus screening out "unique" expressions.

3. The phrases would be normalized, i.e., of the same structure as phrases extracted by the semantic-unit identification operation.

The basic processing steps of the Knowledge Base Building (KBB) Tool illustrated in Figure 2 can be described as follows:

1. Input text is selected. Generally, this will be the titles and abstracts of a large set (150 to 1000) of document records indexed to, or otherwise identified as being related to, a single thesaurus concept. At NASA, a standard online RECON search identifies an accurate set of such records.

2. The text is copied into a file and preprocessed using a simple text-breaking method similar to that used for delineating text phrases for MAI, described earlier.

3. A concatenation process is then used to identify all possible multiword phrases within a maximum length along with certain rules that provide syntactic filtering (which, for example, prevent prepositions and articles from beginning or ending a phrase).

4. A count of the frequency-of-occurrence is determined for each unique single word and multiword phrase. The words and phrases are then sorted in descending order by the frequency values. A lower-limit value is established and phrase units with fewer occurrences than that value are eliminated. A natural bias exists for single-word phrases to occur much more frequently than two-word units, for two-word units to exceed three-word phrases, and so on. This can be dealt with in two ways. A simple

INPUT	PROCESSING	OUTPUT
TEXT FIELDS FROM A CONCEPT-SPECIFIC SEARCH OF THE NASA STI DATABASE	TEXT-BREAKING ROUTINE WORD CONCATENATION PROCESS FREQUENCY SORT PHRASE FILTERING (KB LOOK-UP, NORMALIZATION)	WORD AND PHRASE LISTS CONTAINING FREQUENTLY OCCURRING NATURAL LANGUAGE 'SYNONYMS'

Figure 2. Knowledge Base Building Tool

method is to produce five separate sorts, each one corresponding to a different phrase length. The other method is to use a derived frequency value that effectively accounts for the bias. A process for determining such a value was described by Jones, Gassie, and Radhakrishnan.[5] The formula can be stated as $W \times F \times N^2$, where W is the sum of the frequencies of the words in the phrase, F equals the frequency of the phrase, and N equals the number of distinct words in a phrase.

5. The final processing procedure further refines the output. The phrases are checked against the existing KB entries to eliminate any phrase that properly translates to a thesaurus concept other than the one that the KBB is currently analyzing, and to eliminate single words or multiword phrases identified as having a poor or low semantic value.

Figure 3 shows a sample output from the Knowledge Base Building Tool (KBB). The input consisted of titles and abstracts from records associated with the thesaurus concept METAL MATRIX COMPOSITES. The first column in this figure lists the unedited three-word phrase output. Those phrases selected by a subject analyst for inclusion in the KB are indicated with asterisks. The second column lists the output that the KBB program identifies as being single words. Several acronyms and material abbreviations have been recognized and flagged by a subject analyst.

Single-Word Assessment Tool

One early problem with NASA's MAI system was the preponderance of single-word thesaurus terms that the system generated. About 40 percent of NASA's The-

Un-edited KBB Output for METAL MATRIX COMPOSITES	
THREE-WORD PHRASE OUTPUT	SINGLE WORD OUTPUT
482 * METAL MATRIX COMPOSITE(S)	74 FIBER-MATRIX
72 BEHAVIOR OF COMPOSITES	70 * MMCS
72 STRENGTH OF COMPOSITE(S)	47 REINFORCEMENTS
62 * REINFORCED METAL MATRIX	45 FIBER / MATRIX
61 * ALUMINUM MATRIX COMPOSITE(S)	41 * SIC / AL
55 PROPERTIES OF COMPOSITES	29 STRENGTHENING
51 REINFORCED MATRIX COMPOSITE(S)	29 UNREINFORCED
49 * REINFORCED METAL COMPOSITE(S)	27 * BORON / ALUMINUM
48 * REINFORCED ALUMINUM COMPOSITE(S)	25 MODULI
46 FIBER AND MATRIX	24 * GRAPHITE / ALUMINUM
42 * FIBER REINFORCED METAL(S)	21 * AL-SIC
40 FIBER MATRIX COMPOSITE(S)	20 FP
38 BEHAVIOR OF MATRIX	19 STRENGTHENED
37 BEHAVIOR OF METAL	18 MICROGRAPHS
35 FIBER REINFORCED MATRIX	17 * AL-MATRIX
33 * FIBER METAL COMPOSITE(S)	17 * ARALL
33 * FIBER REINFORCED ALUMINUM	16 ADDITIONS
33 PROPERTIES OF REINFORCED	16 EXTRUDED
32 PROPERTIES OF MATRIX	16 FRACTOGRAPHIC
31 * FIBER METAL MATRIX	16 * GR / AL
30 PROPERTIES OF METAL	16 * GR / MG
29 * ALUMINUM ALLOY MATRIX	16 PARTICULATE-REINFORCED
29 FIBER VOLUME FRACTION	16 SIC-REINFORCED
29 STRENGTH OF FIBER(S)	15 * AL-SI
28 * ALLOY MATRIX COMPOSITE(S)	14 * AL / SIC
28 * ALUMINUM ALLOY COMPOSITE(S)	
28 CHARACTERISTICS OF COMPOSITE(S)	
28 PROPERTIES OF ALUMINUM	
28 PROPERTIES OF FIBER(S)	
27 * METAL MATRIX MATERIAL(S)	
26 * SILICON CARBIDE ALUMINUM	
24 PROPERTIES OF ALLOY(S)	
24 * SIC REINFORCED ALUMINUM	
23 * ALUMINUM METAL MATRIX	
22 BEHAVIOR OF ALUMINUM	
22 HIGH TEMPERATURE COMPOSITES	
22 * REINFORCED ALUMINUM ALLOY(S)	
22 SILICON CARBIDE WHISKER(S)	
22 TRANSMISSION ELECTRON MICROSCOPY	
21 * FIBER ALUMINUM COMPOSITE(S)	
21 THERMAL EXPANSION COEFFICIENT(S)	
20 * CARBIDE REINFORCED ALUMINUM	
20 FATIGUE CRACK GROWTH	

Figure 3. Sample Output from the KBB for METAL MATRIX COMPOSITES

saurus terms are single-word terms; however, indexers in the NASA environment tend to use single-word terms only about 20 percent of the time. In an aerospace database, the term AIRCRAFT, for example, is too general for helpful indexing. An assessment tool was designed to improve the computer-generated set of index terms by reducing the number of single-word terms inappropriately suggested by MAI. A procedure was developed to identify the single-word thesaurus terms that occur frequently in text but are seldom used by human indexers. The terms were sorted and listed by percentages that indicated how often indexers assign one-word terms that appeared in titles and abstracts. This information allowed a close analysis of the problematic single-word terms. Such terms were reviewed, and many were permitted to be translated by the KB only as part of a longer semantic unit.

Maintenance Tools

Statistical text analysis procedures such as the KBB are interesting in that an analysis of their output suggests many possible alternate applications. One problem associated with MAI database maintenance arises because the conceptual domain, as represented by the controlled vocabulary, is dynamic. New terms are added regularly and integrated into existing conceptual hierarchies. Each change in the controlled vocabulary requires at least one change to, addition to, or deletion from, the MAI knowledge base. A single change in the thesaurus can cause numerous knowledge base changes. The areas of the knowledge base requiring modification in response to such changes cannot be easily inferred—particularly if the file happens to be very large. A printed version of the database sorted by the controlled vocabulary terms can help, but it is unwieldy to handle, time-consuming to use, and hard to read; thus, it quickly becomes outmoded.

A modification of the KBB program, the Knowledge Base Maintenance (KBM) program, provides a tool for the identification of the affected entries. The KBM program is essentially the same as the KBB program except that the KBM final procedure allows phrases already translating to a thesaurus term to be included in the output and flagged with two asterisks for easy recognition. Text expressions that have been mapped to an existing thesaurus term in lieu of the newly established term are evident at little more than a glance. The output format of the KBM routine (illustrated below with a few lines of output related to an analysis of the proposed new term MARS CRATERS) includes the ranked phrases with the corresponding thesaurus terms.

Phrase	Term(s)
** 250 CRATERS ON MARS	PLANETARY CRATERS
108 CERULLI CRATER	—
** 102 MARTIAN IMPACT BASIN(S)	MARS (PLANET), STRUCTURAL BASINS

In this case, the KB entries corresponding to the computer-extracted phrases craters on Mars, and Martian impact Basin(s) would be updated by replacing the obsolete term translations MARS (PLANET) and PLANETARY CRATERS with the new term MARS CRATERS. Cerulli crater, when verified as a crater on Mars, would be added as a new entry to be translated to the new term MARS CRATERS.

OTHER APPLICATIONS AND FUTURE DIRECTIONS

The KBM program has also been used as an aid to thesaurus construction. For example, the existing thesaurus terms associated with the identified phrases suggest probable hierarchy locations and related terms for new thesaurus entries. A more interesting application is the use of the phrase output as an aid to identifying trends in the lexicon of a given subject area. It can be particularly useful when investigating an emerging technology or discipline. Sample phrase output for the general area of ROBOTICS is shown in the first column of Figure 4. The second column presents selected output phrases that have been conceptually organized by a subject analyst. Phrases related to the general topical areas within robotics research (dynamics and motion, sensing, vision, control); general system elements (robot arms, joint, end effectors, sensors); and some functional classes (industrial, space, autonomous robots) are represented in the output.

The text input to the KBM is identified through a traditional online search; therefore, the conceptual scope of that text can be easily modified. A separate collection of input text associated with the more specific area of ROBOT VISION was processed in conjunction with the analysis of ROBOTICS. A sample of the output of two-word phrases from the set of documents on ROBOT VISION is shown in Figure 5. The equals (=) signs indicate phrases that are synonymous with the term being analyzed and are therefore candidate "Use" references for the thesaurus. The asterisk (*) signs identify terms that could be included in the thesaurus as hierarchical or related terms.

The limitation of the type of text-based MAI system described above is that the semantic unit is restricted to the level of the phrase. These systems are best suited to document indexing applications associated with science and technology, where the likelihood of a phrase containing specific indexable content is significant. In such environments they can provide a variety of both quality and production benefits.

Some of the possibilities that exist for the future applications of MAI include its use in hypertext search routines, integration into more complex semantic networks, and metathesauri. MAI might also be used to suggest nonhypertext search strategies from natural language inquiries.

In a full-text environment, the basic MAI design could be modified to capture occurrence frequencies of MAI-suggested thesaurus terms. A term weighting scheme could be developed that incorporated these statistical values and special weight values assigned to terms originating from key structural elements of the document, such as title, section headings, abstracts, etc. A carefully crafted system of this kind might even allow its use as a totally automatic database indexing system, producing quality equal to an index constructed manually.

Un-edited KBM Output from General Text on ROBOTICS (Two-word Phrases)		Select KBM Output Phrases Conceptually Organized
172	ROBOT MANIPULATOR(S)	ROBOT DYNAMICS
141	ROBOT ARM(S)	ROBOT MOTION
110	ROBOTIC MANIPULATOR(s)	MANIPULATOR DYNAMICS
99	ROBOTIC SYSTEM(S)	.. TRAJECTORY PLANNING
91	ROBOT CONTROL	.. PATH PLANNING
90	MOBILE ROBOT(S)	.. MOTION PLANNING
84	ROBOT SYSTEM(S)	.. DYNAMIC CONTROL
67	CONTROL SCHEME(S)	.. POSITION CONTROL
65	END EFFECTOR(S)	.. OBSTACLE AVOIDANCE
63	CONTROL ALGORITHM(S)	.. INVERSE KINEMATICS
61	MANIPULATOR SYSTEM(S)	
55	CONTROL LAW(S)	ROBOT JOINTS
53	COMPUTER SIMULATION(S)	
52	INVERSE KINEMATIC(S)	ROBOT ARMS
49	CONTROL ROBOT(S)	MANIPULATOR ARMS
49	VISION SYSTEM(S)	ROBOTIC MANIPULATORS
47	FORCE CONTROL	
46	FLEXIBLE ARM(S)	ROBOT HANDS
44	CONTROL PROBLEM(S)	END EFFECTORS
44	CONTROL STRATEGY(IES)	
44	SPACE ROBOTIC(S)	ROBOT SENSORS
42	AUTOMATION ROBOTICS	.. TACTILE SENSORS
41	AUTONOMOUS ROBOT(S)	.. TACTILE SENSING
40	MANIPULATOR ARM(S)	.. TORQUE SENSORS
39	MANIPULATOR CONTROL	.. FORCE CONTROL
37	DYNAMIC ROBOT	
37	ROBOT DYNAMICS	ROBOT VISION
35	CONTROL METHOD(S)	MACHINE VISION
35	DYNAMIC MODEL	COMPUTER VISION
35	MOTION CONTROL	VISION SYSTEMS
35	TELEROBOTIC SYSTEM(S)	
34	FLEXIBLE MANIPULATOR(S)	ROBOT CONTROL
33	AUTONOMOUS VEHICLE(S)	
32	ADAPTIVE CONTROLLER(S)	ROBOTS
32	FEEDBACK CONTROL	.. AUTONOMOUS ROBOT(S)
32	INDUSTRIAL ROBOT(S)	.. INDUSTRIAL ROBOTS
31	PATH PLANNING	.. SPACE MANIPULATORS
30	CONTROL MANIPULATIORS	.. SPACE TELEROBOTICS
30	DYNAMIC MANIPULATOR(S)	
30	ROBOTIC APPLICATIONS	
30	SPACE TELEROBOTIC(S)	
29	POSITION CONTROL	
29	REDUNDANT MANIPULATOR(S)	
28	CONTROL CONTROL	
27	INTELLIGENT ROBOT(S)	
27	ROBOT MOTION	

Figure 4. Sample Phrase Output for ROBOTICS

Un-edited Output from Text on ROBOT VISION
(Two-word Phrases)

```
205  =  COMPUTER VISION
191     VISION SYSTEM(S)
 60  =  MACHINE VISION
 54  *  OBJECT RECOGNITION
 46  =  ROBOT VISION
 38  *  IMAGE ANALYSIS
 37     ROBOTIC SYSTEM(S)
 35  *  PATTERN RECOGNITION
 33     ROBOT SYSTEM(S)
 33  *  VISION ALGORITHM(S)
 30  *  FEATURE EXTRACTION
 30  *  SCENE ANALYSIS
 28  *  EDGE DETECTION
 27  *  STEREO VISION
 27     VISUAL SYSTEM(S)
 23     RANGE DATA
 22  *  IMAGE SEGMENTATION
 22     MOBILE ROBOT
 22  =  ROBOTIC VISION
 22     THREE-DIMENSIONAL OBJECTS
 20     IMAGE DATA
 20     RECOGNITION SYSTEM
 19  *  LOW-LEVEL VISION
 18  *  IMAGE FEATURES
 18     SPACE APPLICATIONS
 18  *  SURFACE ORIENTATION
 18     VISION APPLICATIONS
```

Figure 5. Sample Output of Two-Word Sematic Units for ROBOT VISION

NOTES

1. Work was performed for the NASA STI Program under contracts NASw-4584, NASw-3330, and NASw-4070. Portions of this chapter are based on the following paper: Genaurdi, M.T. "Knowledge-Based Machine Indexing from Natural Language Text: Knowledge Base Design, Development and Maintenance." In: H. Czap & W. Nedobity (eds.), *TKE '90: Terminology and Knowledge Engineering*, Volume 1. Proceedings of the Second International Congress on Terminology and Knowledge Engineering, 345-351. Frankfurt/M., Federal Republic of Germany: Indeks Verlag. Reprinted as NASA CR-4523.

2. Silvester, J.P. & Klingbiel, P.H. "An Operational System For Subject Switching Between Controlled Vocabularies." *Information Processing and Management*, 29(1), 47-59.

3. Silvester, J.P., Newton, R. & Klingbiel, P.H. "An Operational System for Subject Switching Between Controlled Vocabularies: A Computational Linguistics Approach." (NASA Contractor Report No. 3838). Washington, DC: National Aeronautics and Space Administration. (NTIS No. N85-11903)

4. Klingbiel, P.H. "Phrase Structure Systems in Information Retrieval." *Information Processing and Management*, 21(2), 113-126.

5. Jones, L.P., Gassie, E.W. & Radhakrishnan, S. "INDEX: The Statistical Basis for an Automatic Conceptual Phrase-Indexing System." *Journal of the American Society for Information Science*, 41(2), 87-97.

APPENDIX A

Stopwords

People experienced with using online retrieval systems will generally agree that stopword lists should be small; however, with DTIC's MAI system, the forerunner of NASA's system, stopwords served a different purpose. They functioned, in part, like the stopword lists of KWIC and full text indexing, i.e., words that did not contribute to the formation of the noun phrases sought for indexing were eliminated—or, in this case, made stopwords. For the most part, stopwords were verbs, "weak" adjectives, acronyms, nonsubstantive nouns, adverbs, articles, many prepositions, and conjunctions. During Klingbiel's employment at DTIC, he analyzed a body of four million words of text and determined that nearly half of the words qualified as stopwords because they did not contribute to the formation of indexing phrases.

When Klingbiel established DTIC's second generation system, a problem arose because NASA had authorized index terms that contained verbs, articles, prepositions, and conjunctions, for example, DISRUPTING, A STARS, LOGISTICS OVER THE SHORE (LOTS) CARRIER, and COMMAND AND CONTROL. It was necessary to construct special "grammar" rules to parse text selectively and retrieve legitimate NASA Thesaurus terms that were embedded in titles and abstracts. Also words that were persistently and unacceptably ambiguous were made stopwords.

For NASA's third generation MAI system, the stopword list was streamlined. Stopwords still serve to break up long strings of text and eliminate words that do not contribute to conceptual indexing. The new list, however, contains fewer than four-tenths of one percent of the number of words in the original list. The current stopwords were selected from the text of the entire NASA STI Database through an analysis procedure that identified their high rate of occurrence and limited conceptual importance. The identification of noun phrases has been replaced with the identification of indexable concepts, which may be expressed in any relevant combination of words, excluding the present stopword list. This list is not static; it can be changed and, in fact, may soon have several words removed (such as DESIGNED, IMPLEMENTATION, and TESTED). Only one change has been made since this list was instituted. The list on the following two pages shows the current stopwords.

Stopword List

ABOUT
ABOVE
ACCOUNT
ACHIEVED
ACROSS
ADDITIONAL
AFTER
ALLOW
ALLOWS
ALONG
ALSO
ALTHOUGH
AMONG
AN
ANY
APPROPRIATE
APPROXIMATELY
ARBITRARY
ARE
AROUND
AS
ASPECTS
ASSOCIATED
ASSUMED
AVAILABLE
BASIS
BECAUSE
BEEN
BEING
BEST
BETTER
BOTH
BUT
CAN
CARRIED
CAUSED
CERTAIN
CHARACTERIZED
COMPARED
COMPLETE
CONSIDERATION
CONSIDERED
CONSISTS
CONTAINING
CONTAINS
CONVENTIONAL
CORRESPONDING
COULD
DEFINED

DEMONSTRATE
DEMONSTRATED
DESCRIBE
DESCRIBED
DESCRIBES
DESIGNED
DETAINED
DETERMINE
DETERMINED
DETERMINING
DEVELOP
DEVELOPED
DIFFERENT
DIRECTLY
DISCUSSED
DOES
DUE
DURING
E.G.
EACH
EFFICIENT
EFFORTS
EITHER
EMPHASIS
EMPLOYED
ESPECIALLY
ESTABLISHED
EVALUATE
EVALUATED
EXAMINED
EXAMPLE
EXAMPLES
EXISTING
EXPECTED
EXPERIMENTALLY
FEW
FOUND
FULLY
FUNDAMENTAL
FURTHER
GIVEN
GOOD
GREATER
HAD
HAS
HAVE
HAVING
HERE
HOW
HOWEVER

IDENTIFIED
I.E.
IF
IMPLEMENTATION
IMPORTANCE
IMPORTANT
IMPROVE
INCLUDE
INCLUDED
INCLUDES
INCLUDING
INCREASE
INCREASED
INCREASES
INDICATE
INDIVIDUAL
INTEREST
INTO
INTRODUCED
INVESTIGATE
INVESTIGATED
INVOLVED
INVOLVING
IS
ISSUES
IT
ITS
KNOWN
LESS
MADE
MAJOR
MAKE
MAY
MEANS
MORE
MOST
MUCH
MUST
NECESSARY
NEED
NEEDED
NOT
OBJECTIVE
OBSERVED
OBTAIN
OBTAINED
OCCUR
OTHER
OVERALL
PART

PARTICULAR	SELECTED	THUS
PAST	SEVERAL	TOGETHER
PERFORMED	SHOULD	TOWARD
POSSIBLE	SHOW	TYPES
PREDICT	SHOWED	TYPICAL
PREDICTED	SHOWN	UNDERSTANDING
PRELIMINARY	SHOWS	UNIQUE
PRESENCE	SIGNIFICANT	UP
PRESENT	SIGNIFICANTLY	UPON
PRESENTED	SINCE	USED
PRESENTS	SOME	USEFUL
PREVIOUS	STATUS	USES
PREVIOUSLY	STUDIED	USING
PRODUCE	STUDIES	VARIETY
PRODUCED	STUDY	VARIOUS
PROPOSED	SUB	VERSION
PROVIDE	SUCH	VIA
PROVIDED	SUGGESTED	WAS
PROVIDES	SUITABLE	WE
PROVIDING	SUMMARY	WERE
RECENT	TAKEN	WHEN
RELATED	TESTED	WHERE
RELATIVELY	THAN	WHICH
REPORTED	THAT	WHILE
REQUIRED	THEIR	WHOSE
REQUIRES	THEM	WILL
RESPECT	THEN	WITH
RESULT	THERE	WITHIN
RESULTING	THESE	WITHOUT
RESULTS	THEY	WOULD
REVIEWED	THIS	YEARS
RTOP	THOSE	
SAME	THROUGH	

* A NASA acronym for *Research and Technology Objectives and Plans*

APPENDIX B

General Logic For Machine-aided Indexing with Access-2

The following explains how the NASA MAI system works:

1. The computer breaks the text into word strings by stopping at certain punctuation, such as periods, colons, and semicolons, and at any predesignated stopword.

2. These word strings are then examined, from left to right, in five-word segments, beginning with word one and word two. The first word of every word combination is checked against the knowledge base to see if it exists. If it does not, the word is written out for indexer review.

3. If word one followed by word two is found in the knowledge base as a key to an entry, the posting term field of that entry, which contains the equivalent NASA Thesaurus term(s), is read. There are three possibilities:

a. The posting term field may contain 00, in which case these two words will not generate any NASA Thesaurus term.

b. The posting term field may contain one or more thesaurus terms that will be provided to the indexer as suggested indexing terms.

c. The posting term field will contain an asterisk. This causes the computer to look for an additional word within the five-word segment that, when added to the two previous words, will match the key to another record.

4. If word one followed by word two has an asterisk in the posting term field, and this combination followed by word three, or four, or five does not find a matching key in the knowledge base, then the computer adds 999 (which sorts last in the NASA system) in place of the final word, and tries that combination as a key. If that is not found, the final word in the candidate key is dropped, and replaced with 999. This procedure is repeated, if necessary, until the key is reduced to the first word and 999.

5. If word one followed by word two is not found in the Knowledge Base, then word one is looked up with word three.

6. If word one has been tried with each other word in the five-word segment and no 00, asterisk, or thesaurus term is found, the computer looks up word one followed by 999 to see if a thesaurus term is provided for a single word. This may occur for a strong noun that can stand alone.

7. When the process has used or rejected word one, the five-word segment is again measured off, beginning with word two.

8. A used word is "poisoned," i.e., it is stored with a flag appended, until the processing has passed that word. A poisoned/flagged word may not be used again unless an unpoisoned word is added to it.

Chapter **12**

COMPUTERIZED TOOLS
TO SUPPORT DOCUMENT ANALYSTS

Philip J. Smith, Lorraine F. Normore,
Rebecca Denning and Wayne P. Johnson

INTRODUCTION

As Susanne Humphrey notes in Chapter 8, it is likely that human indexing will retain an important role in the functioning of information retrieval systems for the foreseeable future. Consequently, it is important to ask: What role can the computer play in supporting human indexers, either to improve the quality or the efficiency of their activities? To answer either question, one must find out more about the task of indexing.[1] This is the topic of the studies reported in this chapter.

THE PROBLEM AREA

Chemical Abstracts Services (CAS) employs over 300 document analysts. Broadly speaking, their task is to read published documents and to develop descriptive abstracts and index entries for those documents. These abstracts and entries are then entered into the computer and subjected to a number of background edits that both verify the accuracy of the data and add useful information about its usage. Other document analysts from the same subject area then edit the system output, reviewing the work done by the original document analysts and using the results of the background edits to make the final decision about the index entries that enter the database.

Creating this database is a complex cognitive task, so we began by trying to gain insight into its nature. First, two of the authors were given a brief tutorial on how to do document analysis. Following this, a few experienced document analysts and editors were interviewed and informally observed at work. These exercises revealed at least two areas for concern. The simpler area relates to problems in existing computer support systems, such as slow system response time and difficulties in switching between different existing computerized tools. Many of these problems have since been alleviated by the introduction of new hardware and software systems. Indications persisted, however, that areas for concern existed that were not a result of inadequate technological support. Rather, they relate to the intellectual nature of the task (i.e., memory, problem solving, attention). Identifying specific problems in this area required us to gain a deeper knowledge of document analysis. This led to the two formal empirical studies described in this chapter.

The fundamental questions addressed in these two studies were:

1. What kinds of problems do document analysts now face?
2. Can we identify potentially helpful computer-based tools?

DATA GATHERING

Document analysis, as practiced at Chemical Abstracts Service, demands a great deal of knowledge. Each analyst specializes in a tightly defined area of chemistry and trains in CAS standard analysis procedures for almost a year before becoming fully independent. None of the researchers involved in these studies had equivalent scientific knowledge. We looked, therefore, for data gathering methods that allowed analysts to tell us about the kinds of problems that their work entailed. We were fortunate to find two kinds of tasks already available in the normal work environment that could give us data about problems in document analysis. The first of these involved analyzing data generated by routine peer reviews of abstracting and indexing performance. The second kind of data was gathered during the edit phase normally done in the document analysis workflow.

STUDY 1: AN ANALYSIS OF QUALITY CONTROL REVIEWS

Maintaining the high quality of the database is important to CAS. Within a few of the editorial input groups, it was standard practice to do periodic peer reviews of document analysts' work. These quality control reviews serve three functions. First, they give feedback to the original analyst about general areas or specific instances where improvements could be made. Second, they provide information to management that can be used to suggest general areas for quality improvement. Management may, for example, note a common misconception arising in the quality control reviews and choose to inform all analysts of this problem. Third, reviewers themselves may learn things in the process of reviewing others' work.

Method

As part of this peer review process, documents for review are selected by randomly choosing a batch of approximately eight documents from an analyst's monthly production work. This batch is then given to another analyst for review. The reviewer may or may not have more expertise (in terms of the scientific content area or CAS policies) than the original analyst.

Reviewers are given a specific set of categories for evaluation. They are asked to indicate, for instance, whether the abstract is grammatically correct or the correct index entries were chosen. (See the Results section for a brief list of the actual problem areas). In addition, concrete suggestions are made for changing the abstracts, indexing, etc. These are marked at appropriate points by the reviewer.

For this study, 42 quality control reviews previously prepared in one document analysis unit in the Patent Services Department were used as data. The total number of documents was 305. The number of index entries included in this set of documents was 1,950. The data were drawn from reviews of the work of 16 document analysts that were prepared by eight quality control editors. The data were gathered by counting and categorizing the number of times that problems were reported in the printed questions in a given quality control review. The comments made by the quality control editors were noted.

Results

The forms used to collect information for quality control included both checklist data and comments that were generated by the quality control editor. There were two checklists (one for abstracting and one for indexing).

Problem Reports Derived from "Abstracting" Activities

Table 1 presents data on the frequency with which problems occurred during input activities not associated with indexing ("abstracting"-related activities).

Table 1. Problem Areas in "Abstracting"

Problem Area	Number of Times Problems Reported
Sectioning, cross-references	29
Abstract conciseness	26
Grammar	23
Title coverage	19
Claims coverage	10
Technical content	15
Chemical names, formula	12
Keyword coverage	13
Example appropriateness	8
Keywords	8
Keywords contain important terms	7
Novelty coverage	5

The area found to be most difficult was associated with choosing appropriate section and subsection assignments and with making needed cross-references. A second major area of difficulty was in writing clear and concise abstracts. For example: The phrase "is useful to obtain" was replaced with "produces," while the phrase "the group consisting of" was eliminated entirely. Sometimes, the entire abstract was simply rewritten to make things clear.

The evaluations also indicated that analysts sometimes had trouble writing abstracts that were grammatically correct, such as problems with verb form and tense and the use of articles and prepositions. "Its" was used instead of "their", "of mold" was corrected to read "of the mold," and so on. Correct word usage was sometimes also a problem. Observing the usage of "nickel-phosphorus" instead of "nickel and phosphorus," the editor commented, "a hyphen does not take the place of the word 'and'." To keep this in perspective though, it is important to note that even the most prevalent problem occurred only 29 times in the 305 documents included in the sample.

Problem Reports Derived from Indexing Activities

Table 2 reports equivalent data for problems associated with indexing. The categories have been grouped to show related issues.

The table shows that preparing text modifications for controlled vocabulary terms (TMDs) is a major source of difficulty. Problems both with choosing appropriate starting phrases and with developing clear and accurate text descriptions were noted. For example, one editor made the point that, "at agent-type headings, the chemical name should go first in the TMD."

A second major problem area was in choosing the correct term for subject or substance entries and in applying needed subdivisions (limiters, modifiers, qualifiers, etc.). One indexer chose the general subject term "Sludge" when the appropriate entry was "Waste solids." Sometimes indexers left out index entries that the editors thought should be included.

Study 1 Conclusions

The results of this study made it clear that there are identifiable problems in document analysis related to the cognitive processing needed when doing sectioning, indexing and abstracting. However, the first study had included the work from only one small document analysis unit. We wanted to see if these results could be generalized across the entire range of document analysis departments and if the findings of this first study were accessible using a different data gathering method. Therefore, we embarked upon a second study.

STUDY 2: COLLECTING CONCURRENT VERBAL REPORTS

The study of quality control reviews reported above provided insights into the types of problems encountered during document analysis. To gain further insights into how to help document analysts avoid errors, we conducted a second study in which we observed analysts doing editing. This editing of each document entered into the

Table 2. Problem Areas in Indexing

Problem Area	Number of Times Problems Reported
TMD* starting phrases well-chosen	22
TMD's clear, accurate	34
Used correct substance or general subject main entry and first subdivision**	19
Correct general subject entry and first subdivision**	22
Substance entries	8
Novelty-related claim: substances indexed	19
Novelty-related example: substances indexed	6
Observes cross-references	26
Qualifier	4
Other data elements used correctly	13
All indexed substances marked in original	13
Avoids necessary multiple retrievals	7
Uses general headings for specific cases	8
Trade names input correctly	1

* TMD stands for text modification, the short descriptive phrase used in the CA indexes to give information about the context for each index entry as applied to a given document, e.g., the phrase, "chem. cleaning of, of historic buildings" attached to the index entry "Masonry" for one document in the CA 12th Collective Index.

** The entries in these categories are highly related but are drawn from two different questionnaires and could not be simply collapsed.

CA database is a routine check intended to lessen the chance that either common errors (e.g., typographic, grammatical) or possibly misleading data will make their way into the file. Analysts were asked to talk about what they were thinking (to "think aloud") as they performed the edit stage of document analysis.

Protocol Analysis

We collected verbal protocols to have some way of gaining insights into the cognitive processes of these editors. Protocol analysis provides data that allow us to make inferences about the mechanisms and internal structures of cognitive processes that produce the relations between stimulus and response.[2]

Method

Ten individuals, two from each of five document analysis departments, participated in a 1- to 2-hour protocol analysis session. The editing task was used to structure the interviews. Participants were asked to edit some typical documents and to think aloud as they made changes (edits). Two or three members of the project team were present at each of the sessions. When behaviors were unclear, the editor was asked for further information. The sessions were recorded to provide us with the opportunity of accurately quoting the participants and of verifying the impressions arrived at in the course of the interviews.

Results

The recorded interviews were reviewed and notes made on the kinds of activities noted by the analysts. These activities were summarized in the global "task analysis" that follows.

A Task Analysis

This section describes some of the tasks observed during the course of document editing. Quotes are included to provide a flavor for the activities that editors were engaged in and to show the kinds of data used to generate the task list described in this section. (The list below is a representative, but not complete, list of such tasks).

Task 1. Editors often began work on a batch by presorting the documents, putting articles from the same journal together, commenting: "This is sometimes the kind of thing that's useful if you sort them out and do items from the same journal at the same time."

Task 2. To begin the actual analysis, many editors glanced at the original document. They tended to look quickly at the title and author abstracts and to page through the document to locate notes, structures, tables, and index entry markings made by the original analyst. Knowing where to find information can be useful if the editor must look back at the document to verify an entry. Tables and graphics are especially useful if the document is in a language that the editor does not know well. In addition, analysts/editors come to know the characteristics of journals with which they frequently work and are able to use that knowledge to quickly extract information.

Task 3. The title was read to get a feeling for the type of document being edited. The title was checked for typographical errors and to ensure that it had been completely transcribed from the original and was adequate. [In the Patent Services Department, titles are often enhanced (i.e., words added, deleted, or rearranged) to improve access to the information they contain.]

Task 4. The title was compared with section and subsection assignment and section cross-references to assess the likelihood that sectioning was correct. If there was a disagreement, the editor would try to use the abstract to come to some conclusion about this issue. If there was some uncertainty about the sectioning, the editor might also look at reference materials on subject coverage, exclusions and alternate sectioning:

"I'm reading the title to see if it fits with the section. In this case I'm not so sure."

"I'm going to think about when I read it—if I want to keep the section and subsectioning and cross-references that way."

Task 5. The abstract was checked to ensure internal consistency and chemical "sense":

"Does it seem reasonable that you could go from these nitriles to these glycines, which they say you do?"

"Sodium 3? Something's screwy. You can't have a valence of 3 for sodium."

"The alloy does not contain dispersed ceramics . . . alloy composites contain . . ."

"Usually there's something that gives you an idea there's a problem. The abstract doesn't make sense . . ."

Task 6. The editor tried to make the language of the abstract concise. The goal was not simply to save space but to improve readability.

"I'm reading the abstract to see what it's about. If I see wordy phrases . . ., then I fix it as I go."

"I know I'm going to have half of that in there when I'm through."

"I massaged that into decent English . . . and as condensed as possible."

However, this can be difficult:

"That's the compromise we have to make many, many times, deciding what to put in the text. Sometimes we want to include a number of properties but that would make the text so long we have to lump them together somehow."

Task 7. Editors reviewed the grammar of the abstract (and later of the TMDs). They were concerned with the most obvious types of errors including punctuation mistakes, misuse of prepositions, incorrect sentence structure, and inconsistencies in the use of voice or tense in the content of the abstract.

"That's an awkward sentence."

"That sentence looks messy linguistically."

Knowledge about the native language of the original analyst sometimes alerted the editor to look for predictable problems. For example, it was noted that individuals whose native tongue was German may become confused with gerund endings. Similarly, persons who speak languages such as Chinese, Japanese, and Russian often have problems with definite and indefinite articles.

"The word 'the' was left out of the second sentence. When I see that problem I suspect the analyst is from Japan or China."

Task 8. Editors were alert for typographical errors in the abstract as well as in other parts of the document. In sections in which there are line formulas in the text, special care must be taken to ensure that these are accurate. Chemical names also can create special problems since it is possible that the name is correctly spelled but is not the correct chemical name.

Task 9. The editors next turned to keywording. They checked to see that all important topics had been included. They checked that the concepts included occurred in the title or abstract and that cross-referred sections had appropriate entries among the keywords. If there was no prior reference, they might choose to delete the keyword or phrase or to add information to the abstract.

"I'm looking at these keywords to make sure that all the important chemistry is covered there."

Task 10. From the abstract, title, sectioning and keywords, editors developed expectations about what index entries should have been prepared. As the editors proceeded through the document indexing, they "checked off" the index entries against their expectations. If there were unexpected entries or missing entries, they could either add or delete an index entry as appropriate. In such cases, the editors were likely to look in the original document and to identify either what to add or the reasons that data might have been omitted. Anomalies and missing information were found by cross-checking the content of the abstract, index entries, and the original document and by having a high level of expertise in the area of chemistry being described.

"This seems to be a method (Section 9). Cross reference Section 14 (disease). I'll check to see if a disease is in the keywords."

"I expect to see some specific compounds, a class entry, some tissue entries, and a method entry, so I basically have that in mind right now."

Task 11. The editors evaluated the entries provided by the original document analyst. This decision was often based on complex rules that are tightly tied to practice in a given section. An important determinant, for many editors, concerned the question of promoting access to the document: if it was thought that an added index entry would improve access to the material, it was added; if it simply described the document better but did not provide additional access paths, it was not.

"I'm now trying to decide whether to index rape oil or the compound used to synthesize it, so people who want to look for it [can find it]."

"Is this what we call a polemic? Is there really new information?"

"The hardest decision is when to leave something out."

Task 12. The editors checked the input group to ensure that the original analyst chose appropriate indexing terminology. They determined which name was appropriate for the compound in question. If a name match occurred or a structure was provided, they might use that information to check that the correct name was retrieved from the Registry File (of chemicals). If a name match failed, the editors checked to see whether there was some obvious typographical error or whether they could identify another name which was more likely to cause a retrieval. In some cases, the editors then looked up the alternate name in such sources as the Index Guide:

"I'm going to change 'winterware shoe uppers' to 'winter footwear'."

"Here's a common mistake. The [document's] author uses 'organic acid' instead of 'carboxylic acids.' The input editor didn't know [about this labeling problem] and just used the author's term."

Task 13. The text modifications for each index entry were checked to be certain they reflected document content. Other attributes associated with the entry sometimes needed to be changed from forms that were technically correct but that did not reflect their use in the original document.

"This is a bad entry . . . because it didn't say so in the paper."

"I need to make a change in the auxiliary modifier for a substance. It's not been quite accurately described."

Task 14. The information which was considered technically correct was read through for input errors. Even simple spelling/keying errors can have far-reaching consequences if inappropriate retrievals result.

"I'm checking for keying errors."

"There are two 'withs' together here."

"I'm verifying that the structure is correct. There's a mistake. An 'N' was wrongly typed for an 'H'."

"How do tents fit in if they're discussing shoes? Oh, it's a typo. The keyboarder entered 'tents' instead of 'tests'."

"What we need to check with the name matches is that the input name is correct. Sometimes even a slight miskeying can give you a match but the wrong substance for the document."

Study 2 Conclusions

The data collected in both studies reflect a similar set of problems. The Study 2 results, though, caused us to focus more heavily on the abstract and indexing as a source of difficulty for analysts. They also provided valuable insight into the cognitive concerns that underlie document analysis. We will next discuss our views as to the nature of these cognitive processes and their implications for new support systems for document analysis.

COGNITIVE PROCESSES IN DOCUMENT ANALYSIS

Categorizing the Problems

There are a number of areas where problems arise. To propose solutions, however, it is helpful to further categorize these problems.

Typographic Errors—editors corrected misspellings (often flagged by the computer) resulting from keyboarding errors.

Labeling Errors—in many cases, the indexer has correctly identified a concept that needs indexing but has labeled it wrongly. An example includes using the term "treatment" instead of "roasting" in the keywords.

Content Errors—in some cases, the indexer was wrong about the content of the original document. Sometimes a concept was included that was irrelevant to the original document. In evaluating the keyword list "catalyst support silica gel prepn," for instance, an editor commented: "Review of the patent revealed no mention of catalyst supports."

English Language Usage Errors—verb form and tense and articles and prepositions were sometimes incorrectly used. This occurred most frequently in extended textual data elements (abstracts and TMDs).

CAS Language Usage Errors—in addition to the constraints of normal English usage, CAS document analysts need to conform to CAS language usage guidelines. Problems in this area were indicated by instances in which an indexer had the right concept but formatted the index entry in a way that was inappropriate according to CAS policy.

Problems with Clarity/Conciseness—editors often rewrote phrases to make things clearer or less verbose. Sometimes, the entire abstract was simply rewritten to make things clear.

Analysis as a Task

Analysis includes a wide variety of activities. To be effective, analysts must use considerable knowledge about CAS policy and about specific subject areas in chemistry. Furthermore, they must accomplish their task fairly quickly.

Given the difficult and somewhat repetitive nature of document analysis and normal human limitations, analysts will inevitably make errors. Some of these errors will be due to oversight and some to a lack of knowledge or training in an area; others will be simple "slips" resulting from failures of attention or fatigue.[3]

Strategies and Tactics

To achieve acceptable performance standards, analysts develop a number of strategies to make their effort more efficient. They build up models of "what to expect" when a document is on a particular topic. From its title and abstract, they may conclude that the document falls into a particular class, such as "the analysis of biological compounds." Based on this rapid classification of the document, they can then use the knowledge to structure their analysis.

Editors, too, use their internal models for what should be in a document to review the input data for consistency and completeness. Editors may be "bothered because there's nothing biological in the keywords" or because the index entries fail to discuss typical entries for a document on this type of topic. ("There must be some index entries missed. There has to be something they're analyzing.") They use their knowledge of stereotypical topic areas to help them look for missing entries. ("Now I'm satisfied. I found all the basic elements I was looking for.")

To work efficiently, editors also develop "models" of the work of the particular analyst they are editing. These models indicate such things as the analyst's level of expertise, the subject areas where that analyst is highly experienced, and the types of errors that analyst is most likely to make. Such models allow the editors to accept index entries on faith when the content area is outside their primary area of knowledge.

In the current, batch-oriented system at CAS, no systematic assistance exists for the development of such models. If future support systems provided feedback to the analysts about how similar documents have been analyzed in the past, they could offer a greater opportunity to learn and to encourage consistency in the database building process.

Information Access

Although the net effect of training is to provide analysts with a strong internal model of the way to represent chemical information in the CAS databases, they still must refer to other analysts and to printed and online information resources for problem resolution.

It is important that workers have easy access to authority data and other reference sources. This affects both editing and the original analysis. If analysts can check their input directly against existing terminology, they would be less likely to leave difficult cases for the system and the editor to resolve. Because analysts have the time to work with the document and to understand its content, they are likely to be the best resource for resolving ambiguities between CAS terminology and a given document's content.

Our interviewees suggest various other useful aids for possible development. The ideas include a list of headings arranged by section to aid analysts in identifying appropriate vocabulary. Another useful compilation would be a set of samples of controlled vocabulary terms and associated TMDs (text modifications). It would be helpful if more of this information were available online.

Communication among analysts also needs to be supported. Document analysts and editors need the opportunity to learn about new policies and to discuss issues. One editor noted: "People don't have time to sit down and digest [new policies or subject-matter issues] or sit down in a group and discuss things they don't understand."

231

Allowing for Flexibility

The same job may be done in different ways. Analysts' personal styles may differ; different sections have different types of papers; and different source documents require different strategies. Indeed, in some cases, analysts even disagreed regarding the worth of suggested improvements. One example was the idea that permutations of keywords be shown as the actual words rather than number strings. While some analysts liked this idea, others did not. This example emphasizes a general development criterion to consider: a flexible design that allows analysts to configure displays and support functions to suit individual preferences and needs.

DESIGN SOLUTIONS

From a design perspective, three different solutions could be proposed. The first is to develop programs to assist with the current apprenticeship-based approach to analyst training. The second approach involves extending currently available information sources (hardcopy and batch computer processes) so that the information would be easily accessible online. New sources, too, could be integrated into this online reference umbrella. The third approach is to develop more sophisticated online programs to assist analysts automatically.

Approach I: Training

Training activities take substantial staff time and effort in the editorial departments. A small study of hourly labor charges showed that about 11 percent of total time was consumed by training. Because analyst training methods almost exclusively involve apprenticeship, training time is at least doubly expensive because the trainer's time is also lost production time.

It is important, therefore, to consider methods that could lighten some of the training burden. Many of the problems identified in the preceding analysis result from the need to fit the chemist's knowledge into CAS standard vocabulary and syntactic conventions. We already have a record of CAS standard practice in the existing database. The key suggestion here is to try to use the knowledge already implicit in the CAS database to support ongoing abstracting and indexing practices.

We have come to realize that much of the knowledge needed involves rather narrow areas of chemistry and chemical technology. New trainees and individuals taking cross-training must acquire the necessary scientific knowledge before they can perform adequately as document analysts. They expend considerable effort in coming up to speed on how a particular specialty is represented in the CAS database. Access to examples of the existing database gives the best and most current representation of that information.

Knowledge existing in the database could be exploited for training in a number of ways. At the basic level (arranged in terms of effort/expense) is a mechanism designed to keep files of references useful to individuals coming into the area. These could include 1) review articles and articles with good summary sections to aid the analyst in coming up to speed in the area of chemical specialty, and 2) references to documents that represent good standard practice for indexing that field.

At a more sophisticated level are tools that could enable trainees to search the existing database for related documents. Doing online searching in the course of training has benefits beyond simply retrieving sample documents; it also teaches analysts to try to develop for themselves the questions that actual users may ask as they conduct searches. We should remember that the success of document analysis can be assessed in terms of several questions:

1. Will searchers looking for information on a given topic retrieve a particular document when it is relevant (the problem of search recall)?

2. When retrieved, will this relevant document be mixed in with many less relevant documents (the problem of search precision)?

3. Will searchers be able to judge correctly the document's relevance by looking at the CA title, abstract, keywords, and/or index entries?

Using the existing CAS database as an aid to the trainee should help analysts learn to produce document descriptions that support high levels of search recall and precision. A sample scenario illustrating how this could work is provided below.

An Example of a Retrieval-Based Training Tool

In one window of a multiple window display, trainees would be asked to predict the topics of interest that could lead a searcher to want the document being indexed and to input queries that might be used to search for each topic. A search would then be automatically run for each such query and the associated results displayed. (The computer might try to "wisely" sort the set of documents displayed.)

At this point trainees could choose to modify their initial set of queries (based on the relevance of the documents retrieved by the query). Alternatively, they could proceed to index the document. Upon entering a controlled vocabulary term, any associated description available in the CAS *Index Guide* and a menu of standard subdivisions (qualifiers, categories, limiters, and other modifiers) would be displayed in another window, for example:

Waste Gases

    Usage: Gases discharged into the atmosphere from sources other than furnace firing and combustion engines are indexed at this heading.

                 Waste gases from the firing of furnaces are indexed at Flue Gases.

                 Waste gases from engine exhausts are indexed at Exhaust Gases.

Indexers might then enter or select the desired subdivision. At this point they could request to see a list of the descriptive phases (TMDs) commonly associated with this controlled vocabulary term. (Again, the computer might try to "wisely" sort these items.) Alternatively, they could immediately proceed to enter all or part of the TMD.

The computer would then look for existing documents in the database with "similar" index headings. These would be displayed so that the indexers could see if the current document was placed in the right "neighborhood."

In summary, the benefits of such a retrieval-based training tool are that it:

1. Teaches indexers to think about the potential needs of information seekers;
2. Provides indexers with immediate feedback about their efforts; and
3. Allows trainees to learn syntax and wording by providing relevant examples.

Approach II: Analyst's "Toolchest"

In doing physical work, laborers need the tools necessary for carrying out their jobs. Intellectual workers, too, benefit from using tools that support their work; and most of those tools involve information access. Whether the task is of a physical or intellectual nature, a key element is that workers must choose to use a particular tool rather than having the process take place automatically. And, for the choice to be effective, they need to know which tool is suitable for which task. They also need to have access to the tool when needed. Below we suggest possible tools that CAS analysts can use to deal with problems like those cited above.

Online Information Access Tools

When analysts have identified the item that should be indexed, they must decide how to enter the term in the way most suitable for CAS vocabulary. This problem is greater when analysts are working in sections with which they are not highly familiar. Rapid and easy access to a variety of sources would reduce analyst effort in this area and could, in addition, decrease the amount of editing needed later in the process.

Facilitating access to CAS-specific thesauri could lead the original analyst to preferred labels (e.g., to the use of "transition stage" instead of "cracking stage" or to "forpressing," which was corrected by an editor to read "precompacting"). Online access to the CAS hierarchies could also help indexers identify the right level and terminology for indexing.

Recent work at Chemical Abstracts has focussed on improving access to standard internal documents. Following a diary study of reference resource usage, this project is exploring ways to better present the information in standard reference sources and to integrate information across a variety of internal reference sources in order to better serve CAS document analysts.

A Spell Checker

Analysts would like to have access to a spelling detection program online so that errors such as "labrinth" would be detected (possibly under analyst initiation) during input and corrected then, rather than by the batch edits later in the system (as is currently done). Instead of using a spelling error detection program in a batch mode as is done today, an online tool could not only detect misspelled words online but could also propose lexically and syntactically acceptable alternatives for analyst review and selection. A component in such a system could automatically apply CAS standard abbreviations to incoming text so as to regularize this process.

A Grammar Checker

For problems with both general and CAS-specific language usage, an online tool such as those available in more general writing environments could be developed. Like the Writers' Workbench tools developed by Bell Laboratories, this tool could analyze grammatical usage and give feedback about potential problems. It could review any natural language-like text for common phrases that could be expressed more succinctly. It could highlight and possibly even reformulate these phrases to be more in line with CAS standard practice. These suggestions would then be reviewed by the analyst or editor for correctness.

The tool could be particularized to the syntax found in various CAS data elements (e.g., index entries versus an abstract). Many of the problems found by editors relate to the use of syntax that might be linguistically correct but erroneous according to CAS-defined language guidelines. Providing easy access to definitions, to term-specific approved modifiers to a specific index heading (e.g., qualifiers, limiters, modifiers, etc.), and to sample TMDs could be of assistance.

Approach III: Analyst's Assistant

The preceding approach placed the responsibility for knowing and adhering to CAS usage practice solely on the document analyst and editor. Tomorrow's computer systems will be able to act as intelligent assistants to document analysts. We believe that adding intelligence to the support system could be approached from at least the following four perspectives.

Using Natural Language Processing Techniques

CAS has long been interested in ways to use the methods of computational linguistics to ease the effort of indexing. We are considering the use of natural language understanding techniques to generate candidate index entry-subdivision-TMD triplets from natural language phrases, using our standard authority sources to assist in choice of main entry and subdivision assignment. In the long term, we would like to explore the possibility of generating index terms and TMD phrases using natural language techniques and rich knowledge bases with journal title and author abstract as a data source.

Using a Statistical Approach

Standard practice in computer-generated indexing is to provide index term choices using a statistical analysis of the document title and abstract. It may be possible to construct lists of potential indexing terms using probabilities attached to individual words found in the title/abstract and to present these as candidate entry points for analysis.

Using A Frame-Based Approach

Another approach would be to use our knowledge of the structure in the literature and in the database to fashion an indexing assistant as an aid for those "standard" document types that represent sizable proportion of the work in many CAS sections. This is suggested by the "frame-based" approach to assist searching suggested in Smith's earlier work on an intelligent search assistant to the literature on environmental pollution.[4] (It also has strong ties to work by Humphrey.[5]) This approach suggests that a number of predefinable categories exist that are related to specific topics in the chemical literature. It is possible to construct a list of such categories and to allow analysts simply to "fill in" the information given in a particular document. The following example illustrates a sample (partial) "frame":

Pollutant Frame

Pollutant(s) (Potential or Actual):

Source(s) of Pollutant:

Media Containing Pollutant(s):

Method(s) of Removal:

Substances Used in Removal Process:

Other (concepts not covered by the above

classes):

Novelty:

The structure provided by the prespecified slots would prompt the analyst to think in terms of concept categories that aid searchers. A help function could be provided to display definitions and examples for each entry or slot in the frame. Over time, lists of slot-fillers could be accumulated for viewing. The Assistant could then go on to generate the syntax for potential index entries, which the analyst could review and modify, if necessary. Because the system would automatically generate correctly formatted potential entries, problems with English grammar as well as with CAS internal practices could be greatly reduced.

Using the Knowledge in our Internal Systems

As we have explored the CAS internal reference sources, we have developed insights about ways to use the knowledge contained in them to support indexing more actively. Because these sources are linked by common reference to the CAS controlled vocabulary, we believe that we could bring together information pointing to valid index entries, linking cross-referred terms and usage and editing policy to assist with index entry generation. Our goal is to search for and present information intelligently so as to provide immediate strategic information concerning choice of indexable items to analysts doing online input. This function will extend our ability to provide effective online aids for indexing.

CONCLUSIONS

We began with two questions:

1. What kinds of problems do document analysts now face?
2. Can we identify potentially helpful computer-based tools?

Our investigations uncovered several kinds of problems. Some of these derive from the characteristics of the support systems being used and have been alleviated by employing new systems. Other problems are due to basic human information processing limits: people make errors of omission and commission because they are distracted, tired, or inattentive. Other problems reflect the inherent difficulty of the intellectual task of document analysis. We believe that the latter two types of problem could be ameliorated by a different kind of computer support system, one that focuses on checking for possible error conditions (spelling, grammar) and on bringing together information that would support the human decision making activities that are intrinsically part of document analysis. Systems that incorporate knowledge about the information base being built and about the tasks users are trying to do should be better able to support the next generation of knowledge workers than systems commonly used today.

NOTES

1. Kidd, A. (ed.). *Knowledge Acquisition for Expert Systems: A Practical Handbook*. New York: Plenum Press.

2. Ericsson, K. and Simon, H. *Protocol Analysis: Verbal Reports as Data*. Cambridge, MA: MIT Press, 1984.

3. Norman, D. "Categorization of Action Slips. *Psychological Review*, 88, 1981, 1-15.

4. Smith, P.J., Shute, S., Galdes, D. and Chignell, J. "Knowledge-Based Search Tactics for an Intelligent Intermediary System." *ACM Transactions on Information Systems* 1, 1989, 246-270.

5. Humphrey, S. "MedIndEx System: Medical Indexing Expert System." *Information Processing and Management*, 25, 1989, 73-88.

INTRODUCTION

Edie M. Rasmussen

For many years, online bibliographic databases were composed primarily of document surrogates; that is, citations to documents, associated indexing and, frequently, abstracts. Faced with an information need, users attempted to retrieve relevant documents by formulating a query based on controlled vocabulary, on "free text" search terms, or on some combination of the two. *Controlled vocabulary terms* are index words and phrases manually selected for each document from a set of allowed index terms or codes, frequently a thesaurus. *Free text terms* are those words exclusive of a few common stopwords occurring in the subject related fields of the document record, typically the title and the abstract, for which an inverted index has been created.

The relative effectiveness of controlled vocabulary versus free text for information retrieval from electronic bibliographic databases has been studied extensively, beginning with the Cranfield Tests of the 1960s. Sparck Jones [1981a] examined the major conclusions of the Cranfield Tests. First, the tests identified an inverse relationship between recall and precision; that is, a system's capability of retrieving only relative documents (measured by precision) is inversely related to its capability of retrieving all relevant documents (measured by recall). Second, single term index languages were found to be superior to other types. The Cranfield Tests were followed by others, and a review of twenty years of retrieval systems suggests "(1) that artificial indexing languages do not perform strikingly better than natural language; (2) that complex structured descriptions do not perform strikingly better than simple ones . . . ." [Sparck Jones, 1981b, p. 248]. For the most part, these studies on the behavior of indexing systems were carried out in a restricted experimental environment with relatively small databases of document surrogates. In following the history of the controlled vocabulary-free text debate, Svenonius notes that more recent studies using a case study approach in an operational environment give one consistent outcome: "controlled and uncontrolled vocabularies have different properties, and thus behave differently in retrieval." [p. 334]

Driven by improvements in mass storage and processing capabilities, the database industry increasingly produces full-text rather than, or in addition to, bibliographic databases. Newspapers are a striking example; within just a few years, online availability has gone from one or two well-known metropolitan dailies to a large number of metropolitan, regional, and local papers. As of January 1992, Dialog offered full-text of 35 newspapers across the country in its Papers file of 15 million records. Other available databases contain full-text of collections of journals such as general periodicals or medical journals. Full-text databases of legal materials are also common. Nor are such databases limited to commercial vendors—the availability of cheap optical storage has led many organizations to maintain full-text electronic archives of corporate information.

In combining information retrieval with document delivery, full-text databases provide immediate access to information, but they do so at the cost of increased search time and effort. Potential problems for the searcher of online full-text information are excluded material (graphics or certain categories of information in a newspaper, for instance), difficulty in locating relevant portions of a text, increased data entry errors, minimal or nonexistent indexing, and the need to use stringent statements (for example, by specifying proximity or location of search terms) [Basch, 1989]. The uncontrolled nature of the vocabulary, compounded by the sheer volume of text, presents particular problems.

The early experiments comparing free text and controlled vocabulary used a controlled experimental environment: small databases with a minimal amount of text, which was limited to the most significant part of the document—the title and abstract. "Free text" in this context is not synonymous with the "full-text" of current databases. Accordingly, researchers have shown an interest in evaluating the impact of the availability of full-text on retrieval performance.

Tenopir [1985] used the *Harvard Business Review Online (HBRO)* on a commercial database system to compare retrieval performance from full-text with retrieval from document surrogates (title, abstract, and controlled vocabulary). She found that full-text retrieved more relevant documents than any other method, though at about half the precision level; relative recall (based on total retrieval for all the methods used) was 74 percent. Moreover, full-text retrieval contributed a higher percentage of unique articles than the other methods used. Only the full-text method retrieved documents when the document topic was broader than the search topic, when the full-text compensated for deficiencies in the controlled vocabulary, when concepts were implied in the text, and when the full-text contributed synonyms that were not used in the document surrogates.

The relative efficiency of searching medical databases using index terms and full-text was examined by McKinin et al [1991]. One hundred topics in clinical medicine were searched in the full-text file CCLM and MEDIS as well as in an indexed file, MEDLINE. They found that the full-text files retrieved significantly more relevant documents than the indexed file, though the precision of the indexed file searches was significantly higher. Of the relevant items not retrieved in full-text, 33 percent were missed because of the failure of the searcher to anticipate the natural language used and 57 percent because of strategy problems. Only a few of the citations retrieved by full-text alone would have been retrieved by title or abstract alone, suggesting that full-text adds substantially to recall. The contribution of full-text documents to improved recall was also demonstrated by Ro [1988a, 1988b], who further tested the potential of a variety of weight and ranking techniques to improve the precision levels offered by full-text searching.

Perhaps the most controversial study of full-text retrieval was conducted by Blair and Maron [1985, 1990]. They evaluated an operational retrieval system containing almost 40,000 documents (about 350,000 pages of legal text). Their findings—that less than 20 percent of relevant documents were retrieved at a precision level of 75 percent—were disappointing. Two arguments were offered to explain the poor recall performance of full-text for retrieval: the problem of language, in which the number of ways in which a subject can be represented in a document is unlimited and unpredictable; and the problem of database size, which encourages the searcher to use precise strategies in order to reduce output overload (a tendency that Bates [1984] has identified as "the fallacy of the perfect thirty-item online search"). It has been argued that the Blair and Maron study does not reflect the retrieval performance possible from a full-text system because the system used did not include any of the experimental techniques shown by current research to enhance performance [Salton, 1986, 1992]. Nonetheless, the STAIRS software used is typical of operational commercial online systems, and the database used is typical of the vast amount of correspondence and internal and external documentation associated with any large enterprise.

Given the often disappointing results from full-text searching, it has been argued that some form of controlled vocabulary indexing is necessary for full-text databases. Duckitt [1981] offers three arguments for the use of controlled vocabulary in indexing full-text databases: indexing augments the language of the text and makes specific implied concepts; it offers the capacity to retrieve by category; and it adds at least limited syntactical information in the form of term relationships. While indexing provides a means for dealing with the unstructured nature of the language in full-text, searching by controlled vocabulary also eliminates the power provided by the ability to use any word in a document as a search term, which the research summarized above suggests is a strength. Techniques are needed in searching full-text that harness rather than limit its power.

Commercial database vendors have attempted to facilitate full-text retrieval by providing additional features in the command vocabulary. Examples are the ability to search documents by specifying terms in the same paragraph or (in the case of newspapers) in the lead paragraph; the ability to determine the number of occurrences of a search term within a document; and the ability to browse a retrieved document by viewing only windows of text around a search term. Browsing full-text has been the subject of a research study on the development of an interface for content extraction and navigation [Cove and Walsh, 1988]. Other research has considered the ability to search full-text by treating a document's textual elements (sentences, paragraphs, sections) as retrievable units [Al-Hawamdeh et al, 1991]. The use of a searching thesaurus—a vocabulary list not used to index but invoked in the searching phase—has also been studied [Kristensen and Jarvelin, 1990]. Both positional searching and a searching thesaurus have been incorporated in an expert system developed by Gauch and Smith [1989], which attempts to reformulate user queries automatically in order to improve retrieval from full-text.

The chapters in this section approach the problem of handling full-text from three different perspectives. In "Automatic Indexing," **Donna Harman** presents the state-of-the-art in research on indexing using automatic keyword generation from full-text and, based on these techniques, brings together a series of recommendations for structuring a retrieval system in order to maximize performance. She also considers some advanced techniques for selecting multiword indexing terms, expanding queries, and incorporating natural language processing in the indexing process.

Amy Warner, in "The Role of Linguistic Analysis in Full Text Retrieval," focuses on natural language processing. She argues that the linguistic structure and meaning found in full-text is not exploited by Boolean retrieval techniques as offered by the commercial services, nor by the statistical techniques which are becoming more common, and that information retrieval systems could be improved by adding a linguistic component.

Finally, **Brij Masand**, **Stephen S. Smith**, and **David Waltz**, in "Text Based Applications on The Connection Machine," describe the application of a massively parallel computer to implement an automatic classification scheme on news stories and census forms, and to provide interactive access to large full-text databases using natural language queries and relevance feedback techniques to improve performance.

NOTES

Al-Hawamdeh, S. et al. [1991]. "Using Nearest-Neighbour Searching Techniques to Access Full-Text Documents. *Online Review* 15: 173-191.

Basch, R. [1989]. "The Seven Deadly Sins of Full-Text Searching." *Database* 12(4): 15-23.

Bates, M.J. [1984]. "The Fallacy of the Perfect Thirty-Item Online Search." *RQ* 24: 43-50.

Blair, D.C. and Maron, M.E. [1985]. "An Evaluation of Retrieval Effectiveness for a Full-Text Document Retrieval System." *Communications of the ACM* 28(3): 289-299.

Blair D.C. and Maron, M.E. [1990]. "Full-Text Information Retrieval: Further Analysis and Clarification." *Information Processing & Management* 26(3): 437-447.

Cove, J.F. and Walsh, B.C. [1988]. "Online Text Retrieval Via Browsing." *Information Processing & Management* 24: 31-37.

Duckitt, P. [1981]. "The Value of Controlled Indexing Systems in Online Full Text Databases." *5th International Online Information Meeting*, London, 8-10 December 1981. Oxford: Learned Information. 447-453.

Gauch, S. and Smith, J.B. [1989]. "An Expert System for Searching in Full-Text." *Information Processing & Management* 25: 253-263.

Kristensen, J. and Jarvelin, K. [1990]. "The Effectiveness of a Searching Thesaurus in Free-Text

Searching in a Full-Text Database." *International Classification* 17: 77-84.

McKinin, E.J., Sievert, M.E., Johnson, E.D., and Mitchell, J.A. [1991]. "The Medline/Full-Text Research Project." *Journal of the American Society for Information Science* 42(4): 297-307.

Ro, J.S. [1988a]. "An Evaluation of the Applicability of Ranking Algorithms to Improve the Effectiveness of Full-Text Retrieval. I. On the Effectiveness of Full-Text Retrieval." *Journal of the American Society for Information Science* 39: 73-78.

Ro, J.S. [1988b]. "An Evaluation of the Applicability of Ranking Algorithms to Improve the Effectiveness of Full-Text Retrieval. II. On the Effectiveness of Ranking Algorithms on Full-Text Retrieval." *Journal of the American Society for Information Science* 39: 147-160.

Salton, G. [1986]. "Another Look At Automatic Text-Retrieval Systems." *Communications of the ACM* 29(7): 648-656.

Salton, G. [1992]. "The State of Retrieval System Evaluation." *Information Processing & Management* 28(4): 441-449.

Sparck Jones, K. [1981a]. "The Cranfield Tests." In: *Information Retrieval Experiment* (K. Sparck Jones, ed.). London: Butterworths. 256-284.

Sparck Jones, K. [1981b]. "Retrieval System Tests 1958-1978." In: *Information Retrieval Experiment* (K. Sparck Jones, ed.). London: Butterworths. 213-255.

Svenonius, E. [1986]. "Unanswered Questions in the Design of Controlled Vocabularies." *Journal of the American Society for Information Science* 37(5): 331-340.

Tenopir, C. [1985]. "Full Text Database Retrieval Performance." *Online Review* 9: 149-164.

Chapter 13

AUTOMATIC INDEXING

Donna Harman

INTRODUCTION

Vast amounts of text are available online, including text created for electronic access and text designed mainly for traditional publishing. This text is not searchable without the ability to do automatic indexing. Yet the "discovery" that indexing could be adequate using single terms from the text generally surprised the library community. As Cyril Cleverdon reported from the Cranfield project:

> Quite the most astonishing and seemingly inexplicable conclusion that arises from the project is that the single term indexing languages are superior to any other type . . . unless one is prepared to say that the whole test conception is so much at fault that the results are completely distorted, there is no course except to attempt to explain the results which seem to offend against every canon on which we were trained as librarians. [Cleverdon & Keen 1966]

Today we not only accept these results, but base many of the large commercial online systems on this once revolutionary idea. The discovery of automatic indexing coincided with the availability of large computers and created a major interest in automatically indexing and searching text, such as the work done by H.P Luhn [1957] in investigating the use of frequency weights in automatic indexing. Work has continued since then in various research laboratories and has resulted in more sophisticated automatic indexing methods, using single terms and using larger chunks of text (such as phrases).

This chapter was written to serve two separate goals. The first goal is to provide a tutorial on single term indexing of "real-world" text. The next section, therefore, steps through the indexing process, discussing the critical issues that must be resolved during full text indexing in order to provide effective retrieval performance. Most of these issues are straightforward. Poor choices of indexing parameters, however, produce systems that would be considered failures in most applications.

The second goal is to provide some discussion of advances in automatic indexing beyond the simple single-term indexing used in most operational retrieval systems. The final section discusses many of the techniques being investigated and provides references for further reading.

AUTOMATICALLY PRODUCING SIMPLE INDEX TERMS

This section presents a walkthrough of the processing of an online text file to produce a list of index terms that can be used for searching that file. These terms would be placed in an inverted file, or other data structure, and an information search could be made against this index using Boolean retrieval operators to combine the terms. Alternatively some of the more advanced searching methods could use these terms as input to term weighting algorithms that produce ranked output using statistical techniques.

What Constitutes a Record

The first key decision for any indexing is the choice of record boundaries that identify a searchable unit. A record could be defined as an entire book, a chapter in the book, a section in that chapter, or even a paragraph. This decision is critical for effective retrieval, both in the retrieval/display stage and in the search stage. Often this decision is clear-cut. For example if the application is searching bibliographic records as in an online catalog, clearly a record is one of the bibliographic records. Similarly, if the application is searching newspaper articles or newswire stories for particular events, then each article or story becomes a record.

The choice of record size becomes fuzzy, however, as the size of the documents being examined grows larger. If the documents being searched are lengthy such as legal transcriptions of court cases or full journal articles, then the record might still be the entire document, although this may make display and searching more difficult. If the documents being searched are manuals or textbooks, however, a record should not be the entire document. Here the choice should depend on the retrieval and display mechanisms of the particular application. For example the application of searching an online manual might have a record defined as the lowest subsection, so that users find and display very exact subsections of material. If the application is to provide pointers into paper copies of long articles (such as court cases of 100 pages or more), it might be reasonable to make each page or small section a record so that the display could show a one-line sentence with the hits, and give the page number.

The record size is not only important for display, but also for effective searching. A record which is too short provides little text for the searching algorithms to use, causing poor results. Too large a record, however, may dilute the importance of word matches and cause many false matches. For these reasons it would not be sensible to define a sentence as a record, but a paragraph might be fine. Alternatively it would not be effective to make a very long section a record; it would be better to break it into smaller subsections. Further, the choice of record size may also affect the choice of term weighting and retrieval algorithms (see "Term Weighting" below).

A recent paper [Harman & Candela 1990] shows some possible record size decisions and their consequences. Three different text collections were involved in user testing of a retrieval system using automatic indexing and statistical ranking. The first text collection was small (1.6 megabytes) and consisted of a manual organized into sections and chapters. A record was determined to be equivalent to a paragraph in this manual, because this record size appeared to be the most useful for the end users. This decision caused many short records (see Table 1). The second text collection was a legal code book, with sections and subsections. Here the records were set to be each subsection, again based on user preference. The records were therefore much larger, with many words occurring multiple times within each record. The third text collection consisted of about 40,000 court cases. A record here was set to be a court case.

Table 1 shows some basic statistics on these text collections. The average number of terms per record includes duplicate terms and is a measure of the record length rather than the number of unique term occurrences. The average number of postings per term is the average number of documents containing that term.

TABLE 1 Collection Statistics			
Size of collection	1.6 MB	50 MB	806 MB
Number of records	2653	6652	38304
Average number of terms per record	96	1124	3264
Number of unique terms	5123	25129	243470
Average postings per term	14	40	88

What Constitutes a Word and What "Words" to Index

The second key decision for any indexing is the choice of what constitutes a word and, then, which of these words to index. In manual indexing systems this choice is easily made by the human indexer. For automatic indexing, however, it is necessary to define what punctuation should be used as word separators and to define what "words" to index.

Normally word separators include all white spaces and all punctuation. But there are many exceptions to this rule and, depending on the application and the searching software, the methods of handling these exceptions can be crucial to successful retrieval. The following examples illustrate some of the problems encountered in typical applications.

Hyphens—some words can appear in both hyphenated and unhyphenated versions. Sometimes the treatment of hyphens is critical to retrieval, such as in chemical names and other normally hyphenated elements (glycol-sebacic, F-15, MS-DOS, etc.).

Periods—periods can appear as a part of a word, such as computer file names (paper.version1), subsection titles (1.367A), and in company names.

Slashes, parentheses, underscores—these can appear as parts of words (OS/2), as parts of section titles (367(A)), and as parts of terms in programming languages (doc_no).

Commas—if numbers are indexed, commas and decimal points become important.

Once word boundaries are defined, an equally difficult issue is what words or tokens to index. This applies particularly to the indexing of numbers. If numbers are indexed, the amount of unique words can explode because an unlimited set of unique numbers exist. As an example, when all numbers in the 50-megabyte text collection shown in Table 1 were indexed, the number of unique terms rose from 10,122 (indexing no numbers) to 55,486 (indexing all numbers). (The amount of unique terms shown in Table 1 includes indexing some numbers as explained later). Indexing all numbers would have nearly doubled the index size, and therefore caused slower response times. Not indexing the numbers, however, can lead to major searching problems when a number is critical to the query (such as "what were the major breakthroughs in computer speed in 1986").

The same problem can apply to the indexing of single characters (other than the words "a" or "i", which are discussed in the next section as stopwords). Whereas the number of unique single characters are limited, the heavy use of single characters as initials, section labels, etc. can increase the size of the index. Again, however, not indexing single characters can lead to searching problems for queries in which these characters are critical (such as "sources of vitamin C").

The solutions to both the problem of word boundaries and what words to index involve compromises. Before indexing is started, samples of the text to be indexed, and samples of the types of queries to be run, need to be closely examined. This may require a prototype/user testing operation, or may be solved by simply discussing the problem with the users. The following examples illustrate some of the possible compromises.

• The punctuation in the text should be studied, and potential problems identified so that reasonable rules of word separation can be found. Often hyphenated words are treated both as separated words and as hyphenated words. Other types of punctuation are handled differently based on preceding or succeeding characters or spaces.

• The use of upper and lower case letters also needs to be determined. Usually upper-case letters are changed to lower case during indexing as words capitalized in sentence beginnings will not correctly match lower case query words. If proper nouns are to be treated as special terms, however, their recognition requires upper-case letters.

• The indexing of numbers is also heavily application dependent. Dates, section labels, and numbers combined with alphabetics may be indexed, and other numbers not indexed. If hyphens can be kept, then some number problems are eliminated (such as F-15). In the 50-megabyte text collection shown in Table 1, numbers that were part of section labels were kept, and these were distinguished by the punctuation that appeared in the number. Some searches were still unsuccessful, however, due to the lack of complete number indexing.

• The indexing of single characters is somewhat easier to handle. Potential users can check the alphabet and note any letters that have particular meaning in their application, and these letters can be indexed.

Most commercial systems take a conservative approach to these problems. For example, Chemical Abstracts Service, ORBIT Search Service, and Mead Data Central's LEXIS/NEXIS systems all recognize numbers and words containing digits as index terms, and all are case insensitive. In general they have no special provisions for punctuation marks, although Chemical Abstracts Service keeps hyphenated words as single tokens, and the other two systems break hyphenated words apart [Fox 1992].

Use of Stop Lists

Additionally most automatic indexing techniques work with a stop list that prevents certain high-frequency or "fluff" words from being indexed. Francis & Kucera [1982] found that the ten most frequently used words in the English language typically account for twenty to thirty percent of the terms in a document. These terms use large amounts of index storage and cause poor matches (although this is not usually a problem because of the use of multiple query terms for matching purposes).

One commonly-used approach to building a stop list is to use one of the many lists generated in the past. Francis & Kucera [1982] produced a stop list of 425 words derived from the Brown corpus, and a list of 250 stopwords was published by van Rijsbergen [1975]. These lists contain many of the words that always have a high frequency, such as "a", "and", "the", and "is", but also may contain "fluff" words that may not have a high frequency for some text collections, such as "below", "near", "always", and "that." Note that unlike high frequency words, "fluff" words do not necessarily hurt retrieval performance, and will not seriously affect storage. Often these words become crucial to retrieval, such as in a query "stocks with costs below X dollars," or "restaurants near the harbor."

A more suitable method of constructing a stop list would be to produce a word frequency listing for the text to be indexed, and then examine each of the high frequency words. If a given word in the application has no known importance, then that word can be safely placed on a stop list. An example of this procedure is the work done at the National Institute of Standards and Technology (NIST) with a 250-megabyte collection of *The Wall Street Journal*. The top 27 high-frequency words were examined, and 4 words were removed as possibly important ("a", "at", "from" and "to"). The remaining twenty-three words then became the stop list. This was a reduction from a previously used stop list from the SMART project of 418 words. The shrinkage of the stop list caused an increase of about 25 percent in the index storage, but made available for searching an additional 395 words. This new stop list is shown as Table 2 to illustrate an abbreviated stop list rather than to recommend any particular one.

TABLE 2 Sample Stop Words				
an	been	in	or	which
and	but	is	that	will
are	by	it	the	with
as	for	of	this	
be	have	on	was	

It should be noted that commercial systems are even more conservative in the use of stop lists. ORBIT Search Service has only eight stop words: "and", "an", "by", "from", "of", "or", "the", and "with" [Fox 1992]. The MEDLARS system has even fewer stop words.

Use of Suffixing or Stemming

Many information retrieval systems also use suffixing or stemming to replace all indexed words with their root forms. Different stemming algorithms have been used, including "standard" algorithms, and algorithms built for a specific domain such as medical English [Pacak 1978]. For a survey of the various algorithms see Frakes [1992]. Three standard algorithms, an "S" stemming algorithm, the Lovins [1968] algorithm, and the Porter [1980] algorithm, are most often used, and the following excerpts [Harman 1991] show some of their characteristics.

The "S" stemming algorithm, a basic algorithm conflating singular and plural word forms, is commonly used for minimal stemming. The rules for a version of this stemmer, shown in Table 3, are only applied to words of sufficient length (three or more characters), and are applied in an order dependent manner (i.e., the first applicable rule encountered is the only one used). Each rule has three parts: a specification of the qualifying word ending, such as "ies"; a list of exceptions; and the necessary action.

TABLE 3 An "S" Stemmer

IF   a word ends in "ies", but not "eies" or "aies"
     THEN "ies" --> "y"
 	
 IF   a word ends in "es", but not "aes", "ees", or "oes"
     THEN "es" --> "e"
 	
 IF   a word ends in "s", but not "us" or "ss"
     THEN "s" --> NULL

The Lovins stemmer works similarly, but on a much larger scale. It contains a list of over 260 possible suffixes, a large exception list, and many cleanup rules. In contrast, the Porter algorithm looks for about 60 suffixes, producing word variant conflation intermediate between a simple singular-plural technique and Lovins algorithm. Table 4 shows an example of the differences among the three stemmers. The first column shows the actual words (full words) from the query. The next three columns show the words that are conflated with the original words (words that stem to the same root for that stemmer) based on three different stemmers. The starred terms are the ones that were useful in retrieval for this particular query and are shown only to indicate the "quasi-random" matching that occurs when matching query terms with terms in relevant documents.

TABLE 4			
Stemmer Differences for query 109 of the Cranfield test collection			
Query -- panels subjected to aerodynamic heating			
FULL WORD	S	PORTER	LOVINS
*panels	*panel *panels	*panel *panels	*panel *panels
subjected	subjected	subjected *subject subjective subjects	subjected *subject subjective subjects
*aerodynamic	*aerodynamic aerodynamics	*aerodynamic aerodynamics *aerodynamically	*aerodynamic aerodynamics *aerodynamically aerodynamicist
*heating	*heating	*heating *heated	*heating *heated *heat heats heater

Stemming or suffixing is done for two principal reasons: the reduction in index storage required and the increase in performance due to the use of word variants. The storage savings using stemming is data and implementation dependent. For small text collections on machines with little storage, a sizable amount of inverted file storage can be saved using stemming. For the 1.6-megabyte manual shown in Table 1, approximately 20 percent of storage was saved by using the Lovins stemmer. Lennon et al [1981] showed compression percentages for the Lovins stemmer of 45.8 percent for the Brown corpus. However, for the larger text collections normally used in online retrieval, less storage is saved. The savings was less than 14 percent for the text of 50 megabytes in Table 1, probably because this text contains large amounts of numbers, misspellings, proper names, etc. (items that usually cannot be stemmed).

In terms of performance improvements, research has shown that *on the average* results were not improved by using a stemmer. However, system performance must reflect a user's expectations, and the use of a stemmer (particularly the S stemmer) is intuitive to many users. The OKAPI project [Walker & Jones 1987] did extensive work on improving retrieval in online catalogs, and strongly recommended using a "weak" stemmer at all times, as the "weak" stemmer (removal of plurals, "ed" and "ing") seldom hurt performance but provided significant improvement. They found drops in precision for some queries using a "strong" stemmer (a variation of the Porter algorithm), and therefore recommended the use of a "strong" stemmer only when no matches were found.

One method of selective stemming is the availability of truncation in many online commercial retrieval systems. Frakes [1984] found, however, that automatic stemming performed as well as truncation by an experienced user, and most user studies show little actual use of truncation.

Given today's retrieval speed and the ability for user interaction, a realistic approach for online retrieval would be the automatic use of a stemmer, using an algorithm like Porter or Lovins, but providing the ability to keep a term from being stemmed (the inverse of truncation). If a user found that a term in the stemmed query produced too many nonrelevant documents, the query could be resubmitted with that term marked for no stemming. In this manner, users would have full advantage of stemming, but would be able to improve the results of those queries hurt by stemming.

ADVANCED AUTOMATIC INDEXING TECHNIQUES

The basic index terms produced by the methods discussed in the first section can be used "as is," with Boolean connectors to combine terms, or a single term may be used for simple searches. Researchers in information retrieval have been developing more complex automatic indexing techniques for over 30 years, however, and having varying degrees of success with these new techniques doing experiments with small test collections. Some of these techniques (such as the term weighting discussed next) are clearly successful and scale easily into large full-text documents. Other techniques, such as the query expansion techniques described later, do well on small test collections, but may need additional experimentation when used in large full-text collections. The added discrimination provided by using phrases as indexing terms rather than only single terms is also discussed later. In general the use of phrases has not been successful in small test collections, but is likely to become more useful, or even critical, in large full-text documents. Large full-text collections may need better term discrimination measures, and some recent experiments in selecting better indexing features or in providing more advanced term weighting are described in separate sections. Finally, the notion of combining evidence from multiple types of document indexing is presented.

Term Weighting

Whereas terms coming from automatic indexing can be used without weights, they offer the opportunity to do automatic term weighting. This weighting is essential to all systems doing statistical or probabilistic ranking. Many of the commercial systems provide a capability to rank documents based on the number of terms matching between the query and the document, but find that users do not select this option often due to its poor performance caused by one or more of the following issues:

1. No technique for resolving ties. If a query has three words, it may be that only a few documents match all three words, but many will match two terms, and these documents are essentially unranked with respect to each other.

2. No allowance for word importance within a text collection. A query such as "term weighting in information retrieval" could return a single document containing all four noncommon words, and then an unranked list of documents containing the two words "term" and "weighting" or "information" and "retrieval," all in random order. This could mean that the possibly 10 documents containing "term" and "weighting" are buried in 500 documents containing "information" and "retrieval".

3. No allowance for word importance within a document. Looking again at the query "term weighting in information retrieval," the correct order of the documents containing "term" and "weighting" would be by frequency of "weighting" within a document, so that the highest ranked document contains multiple instances of "weighting," not just a single instance.

4. No allowance for document length. Whereas this factor is not as important as the first three factors, it can be important to normalize ranking for length; otherwise, long documents often rank higher than short documents even though the query terms may be more concentrated in the short documents.

These problems can be largely avoided by using more complex statistical ranking routines involving proper term weighting or accurate similarity measures.

Various experiments have been concerned with developing optimal methods of weighting the terms and optimal methods of measuring the similarity of a document and the query. One of the term weighting measures that has proven very successful is the inverted document frequency weight or IDF [Sparck Jones 1972], which basically measures the scarcity of a term in the text collection. A second measure used is some function of a term's frequency within a record. These measures are often combined, with appropriate normalization factors for length, to form a single term weight. Statistically-ranked retrieval using this type of term weighting has a retrieval performance that is significantly better (in the laboratory) than using no term weighting [Salton & McGill 1983, Croft 1983, Harman 1986].

The following recommendations can be made based on this research:

1. The use of term weighting based on the distribution of a term within a collection usually improves performance, and never hurts performance. The IDF measure has been commonly used for this weighting.

$$IDF_i = \log_2 \frac{N}{n_i} + 1 \quad \textit{(Sparck Jones 1972)}$$

where N = the number of documents in the collection
n_i = the total frequency of term i in the collection

2. The combination of the within-document frequency with the IDF weight often provides even more improvement. It is important to normalize the within-document frequency in some manner, both to moderate the effect of high frequency terms in a document (i.e. a term appearing 20 times is not 20 times as important as one appearing only once) and to compensate for document length. Data containing very short documents (such as titles only) should not use weighting for within-document frequency. The following within-document frequency measures illustrate correct normalization procedures.

$$cfreq_{ij} \;=\; K \;+\; (1-K)\,\frac{freq_{ij}}{maxfreq_j} \qquad (Croft\ 1983)$$

$$nfreq_{ij} \;=\; \frac{\log_2 (freq_{ij}+1)}{\log_2 length_j} \qquad (Harman\ 1986)$$

> where $freq_{ij}$ = the frequency of term i in document j
> $maxfreq_j$ = the maximum frequency of any term in document j
> K = the constant used to adjust for relative importance of within-document frequency
> $length_j$ = the number of unique terms in document j

3. Assuming use of within-document term frequencies, several methods can be used for combining these with the IDF measure. Both the combining of term weighting and the use of this weighting in similarity measures between queries and documents are shown.

$$similarity\,(Q\,,D) \;=\; \frac{\sum\limits_{i=1}^{t}(w_{iq} \times w_{ij})}{\sqrt{\sum\limits_{i=1}^{t}(w_{iq})^2 \times \sum\limits_{i=1}^{t}(w_{ij})^2}} \qquad (Salton\ \&\ Buckley\ 1988)$$

$$where \quad w_{iq} \;=\; (0.5 + \frac{0.5\,freq_{iq}}{maxfreq_q}) \times IDF_i$$

$$and \quad w_{ij} \;=\; \frac{freq_{ij} \times IDF_i}{\sqrt{\sum\limits_{vector}(freq_{ij} \times IDF_i)^2}}$$

> where $freq_{iq}$ = the frequency of term i in query q
> $maxfreq_q$ = the maximum frequency of any term in query q
> IDF_i = the IDF of term i in the entire collection
> $freq_{ij}$ = the frequency of term i in document j

Salton & Buckley suggest reducing the query weighting w_{iq} to only the within-document frequency ($freq_{iq}$) for long queries containing multiple occurrences of terms, and to use only binary weighting of documents (w_{ij} = 1 or 0) for collections with short documents or collections using controlled vocabulary.

$$similarity_j = \sum_{i=1}^{Q} (C + IDF_i \times cfreq_{ij}) \qquad (Croft\ 1983)$$

$$where \quad cfreq_{ij} = K + (1-K)\,\frac{freq_{ij}}{maxfreq_j}$$

where $freq_{ij}$ = the frequency of term i in document j
C = the constant used to adjust for relative importance of all term weighting
$maxfreq_j$ = the maximum frequency of any term in document j
K = the constant used to adjust for relative importance of within-document frequency

C should be set to low values (near 0) for automatically indexed collections, and to higher values such as 1 for manually-indexed collections. K should be set to low values (0.3 was used by Croft) for collections with long (35 or more terms) documents, and to higher values (0.5 or higher) for collections with short documents, reducing the role of within-document frequency.

$$similarity_j = \sum_{i=1}^{Q} \frac{\log_2\,(freq_{ij}+1) \times IDF_i}{\log_2 length_j} \qquad (Harman\ 1986)$$

where $freq_{ij}$ = the frequency of term i in document j
$length_j$ = the number of unique terms in document j

4. It can be very useful to add additional weight for document structure, such as for terms appearing in the title or abstract versus those appearing only in the text. This additional weighting needs to be considered with respect to the particular text collection being used for searching.

This section on term weighting presents only a few of the experimental techniques that have been tried. For a more thorough survey, see Harman [1992a].

Query Expansion

A problem found in all information retrieval systems is that relevant documents are missed because they contain no terms from the query. Whereas users often do not want to find most of the relevant documents, sometimes they want to find many more relevant documents and are willing to examine more documents in hopes of finding more relevant ones. However, the automatic indexing systems generally do not offer the "higher-level" terms describing a document that could have been manually assigned, and it is difficult to generate a more exhaustive search. One way around this difficulty is to provide tools for query expansion. A simple example of such a tool would be the ability to browse the text collection's dictionary or word list. Two more sophisticated techniques would be the use of relevance feedback or the use of a thesaurus constructed automatically.

The relevance feedback technique allows users to select a few relevant documents and then ask the system to use these documents to improve performance, i.e., retrieve more relevant documents. A significant amount of research has gone into using this method, although few user experiments have been done on large test collections. Salton & Buckley [1990a] showed that adding relevance feedback to their similarity measure results in up to 100 percent improvement for small test collections. Croft [1983] used the relevant and nonrelevant documents to change the term weighting probabilistically and, in 1990, he extended this work by also expanding queries using terms in the relevant documents. A similar approach was taken by Harman [1992c] and these results (again for a small test collection) showed performance improvements of around 100 percent. Clearly the use of relevance judgments to improve performance is important in full-text searching and can supplement the use of the basic automatically-indexed terms, but the exact methods of using these relevance judgments for large full-text documents remains undetermined. Possibly their best use is in providing an interactive tool for modifying the query by suggesting new terms. For a survey of the use of relevance feedback in experimental retrieval systems, including Boolean systems, see Harman [1992b].

A different method of query expansion could be the use of a thesaurus. This thesaurus could be used as a browsing tool or incorporated automatically in some manner. The building of such a thesaurus, however, is a massive, often domain-dependent task. Some research has been done into automatically building a thesaurus. Sparck Jones & Jackson [1970] experimented with clustering terms based on co-occurrence of these terms in documents. They tried several different clustering techniques and several different methods of using these clusters on the manually indexed Cranfield collection. The major results on this small test collection showed that it is 1) important NOT to cluster high frequency terms (they became unit clusters); 2) important to create small clusters; and 3) better to search using the clusters alone rather than a "mixed-mode" of clusters and single terms. Crouch [1988] also generated small clusters of low frequency terms, but had good search results using query terms augmented by thesaurus classes. Careful attention was paid to weighting properly these additional "terms." How these results scale up to large full-text collections is, of course, unknown, but the concept seems promising enough to encourage further experimentation.

The Use of Multiple-Word Phrases for Indexing

Large full-text collections need not only special query expansion devices to improve recall (the percentage of total relevant documents retrieved), but also precision devices to improve their accuracy. One important precision device is the term weighting discussed earlier. The ability to provide ranked output improves precision because users are no longer looking at a random ordering of selected documents. However, further improvement in precision may be necessary for searching in large full-text collections, and one way to get additional accuracy is to require more stringent matching, such as phrase matching.

Phrase matching has been used in experiments in information retrieval for many years, but the technique is now getting more attention because of improvements in natural language technology. The initial phrase matching used templates [Weiss

1970] rather than deep natural language parsing algorithms. The FASIT system [Dillon & Gray 1983; Burgin & Dillon 1992] used template matching by creating a dictionary of syntactic category patterns and using this dictionary to locate phrases. They assigned syntactic categories by using a suffix dictionary and exception list. The phrases detected by this system were normalized and then merged into concept groups for the final matching with queries.

A second type of phrase detection method that is based purely on statistics was investigated by Fagan [1987, 1989]. This type of system relies on statistical co-occurrences of terms, as did the automatic thesaurus building described above, but requires that these terms co-occur in more limited domains (such as within paragraphs or within sentences), and within a given proximity of each other. Fagan investigated the use of many different parameters for selecting these phrases, and then added the phrases as supplemental index terms, i.e., all single terms were first indexed and then some additional phrases were produced.

Fagan [1987] also examined the use of complete syntactic parsing to generate phrases. The parser generated syntactic parse trees for each sentence and the phrases were then defined as subtrees of those parse trees that met certain structural criteria. Salton et. al. [1989, 1990a] compared the phrases generated for two book chapters both by the statistical methods and the syntactic methods and found that both methods generated many correct phrases, but that the overlap of those phrases was small. Salton et. al. [1990b] also tried a syntactic tagger and bracketer [Church 1988] to identify phrases. The tagger uses statistical methods to produce syntactic part-of-speech tags, and the bracketer identifies phrases consisting of noun and adjective sequences. This simpler approach does not require the completion of entire parse trees and seemed to produce as many good phrases.

In general, retrieval experiments that add phrases to single term indexing have not been successful with small test collections. One reason has been the scarcity of phrases in the text that match phrases in the query. Lewis & Croft [1990] tried first locating phrases using a chart parser and then clustering these phrases. The retrieval used single terms, phrases, and clustered phrases in different combinations. The best performance used terms, phrases, and clustered phrases as features for retrieval. Even this performance, however, was not significantly better than performance using only single terms for the small test collection used.

The current belief among researchers is that the use of multiple-word phrases will be successful only for large collections of text, partially because of the need for enough text to locate phrases that will be good features for retrieval. Equally important, the higher precision retrieval offered by phrases may only be important in the larger full-text retrieval environment. Croft et. al. [1991] investigated various ways of both generating and using phrases in retrieval, and although their results on the small CACM test collection were not significant, the work they are doing using phrases on a larger test collection shows impressive results. It is likely that the use of phrases for retrieval in large full-text retrieval environments will show significant and, possibly, critical improvements over single term indexing.

Feature Selection

Another method of improving precision in retrieval from large full-text data is to select indexing features more carefully. The current approach to automatic indexing generally indexes all stems in a document, eliminating only stopwords and, possibly, numbers. This exhaustive coverage may be important for small documents such as abstracts or bibliographic records, but using all terms in very large records may weaken the matching criteria. Ideally one would like to be able to select automatically the single terms or phrases that best represent a document—but this area has attracted little research because of the absence of large full-text test collections.

Two recent papers address this issue. The first [Strzalkowski 1992] described some research using a statistical retrieval system with some improvements based on natural language techniques. Strzalkowski used a very fast syntactic parser to parse the text. The phrases found using this parser were then statistically analyzed and filtered to produce automatically a set of semantic relationships between the words and subphrases. This highly selective set of phrases was then used to both expand and filter the query. The results on the small CACM collection showed a significant improvement in performance over the straight statistical methods, and these techniques clearly will scale up to larger full-text documents.

The second paper [Lewis 1992] investigated feature selection using a classification test collection. This test collection contains 21,450 Reuters newswires that have been manually classified into 135 topic descriptions. The goal of this research was to identify what text features (terms, phrases, or phrase clusters) were important in generating these categories. Best results were obtained for a small number of features (10 - 15), and some discussion is made of the best ways to select these features. This type of approach to feature "pruning" also needs further exploration for large full-text collections.

More Advanced Term Weighting Techniques

A third approach to increased precision for the larger documents is to use all terms for indexing, but to provide more sophisticated term weighting methods than those discussed previously. Salton & Buckley [1991] presented results from work using an online encyclopedia in which they weighted terms both globally for an entire document (as described previously), but also locally for a given sentence. In this particular experiment they performed multiple-stage searching in which a short initial query was used to find one or more relevant sections or paragraphs, and these sections were then used to find similar sections using both global and local weighting schemes. Whereas the global weights help increase the recall by returning many similar items, the local weights can be used as a filtering operation to improve the precision of the returned set. Further details can be found in [Salton et al. 1993]. This type of approach to searching and term weighting may be particularly suitable for large full-text data collections.

Using Combinations of Indexing Techniques

All the preceding research efforts are based on combining various information elements from the text to improve indexing and searching. The best term weighting schemes combined different statistical measures of term importance. The section on query expansion dealt with combining information about term co-occurrence to identify automatically better query terms and term weights. The work on multiple word phrases investigated how to locate phrases, but also how to combine these phrases correctly with single terms. Feature selection involves combining information from the text to help better select which features to index, and the advanced term weighting techniques combine term weights at two granularity levels to improve precision.

Other more explicit combination techniques have been tried, from simple user weighting of terms (to be combined with the statistical term weighting), to combining of database attributes with free text [Deogun & Raghavan 1988], to more elaborate combining of concepts such as citations, attributes, and data into the vector space model [Fox et al. 1988]. Results have generally shown improvements in performance, even for small test collections. This combination of various sources of information can be extended to combining various types of indexing (such as manual or automatic), various types of queries (such as using or not using Boolean connectors), or various types of searching (such as cluster searching versus document searching). It has been shown [Katzer et al. 1982] that different indexing or searching methods can produce comparable results, but with *little* overlap between the sets of relevant documents. Clearly it would be ideal to combine these methods, but the method for combining the completely different approaches to indexing and searching is not readily apparent.

A new model, the inference network [Turtle and Croft 1991] is designed specifically for the task of combining evidence or probabilities from all these different methods. This network consists of term nodes, document nodes, and query nodes, connected by links with probabilistic weighting factors, and can be used to try multiple ways of combining information from these nodes to form a list of documents ranked in order of likely relevance to a user's need. Turtle and Croft show how this model can be used to represent most of the basic indexing and searching techniques, and discuss how the generation of this model provides the scope for a thorough investigations of how to perform complex combinations of techniques. This type of representation can be viewed as a quite advanced indexing method, and allows important capabilities in handling large full-text data.

SUMMARY

Whereas the traditional automatic single term indexing described in this chapter enables reasonable searching of large full-text documents, the more advanced techniques discussed may all prove critical in raising the retrieval performance beyond a mediocre level. It is critical that research continue into these advanced techniques and others like them; and that (as they become proven methodologies), they be accepted as standard automatic indexing techniques by the information retrieval community as a whole.

NOTES

Burgin R. and Dillon M. [1992]. "Improving Disambiguation in FASIT." *Journal of the American Society for Information Science*, 43(2), 101-114.

Church K. [1988]. "A Stochastic Part Program and Noun Phrase Parser for Unrestricted Text." In: *Proceedings of the Second Conference on Applied Natural Language Processing, 1988*. 136-143, Austin, Texas.

Cleverdon C.W. and Keen E.M. [1966]. *Factors Determining the Performance of Indexing Systems, Vol. 1: Design, Vol. 2: Test Results*. Aslib Cranfield Research Project, Cranfield, England, 1966.

Croft W.B. [1983]. "Experiments with Representation in a Document Retrieval System." *Information Technology: Research and Development*, 2(1), 1-21.

Croft W.B. and Das R. [1990]. "Experiments with Query Acquisition and Use in Document Retrieval Systems." In: *Proceedings of the 13th International Conference on Research and Development in Information Retrieval*; September 1990, 349-368; Brussels, Belgium.

Croft W.B., Turtle H., and Lewis D. [1991]. "The Use of Phrases and Structured Queries in Information Retrieval." In: *Proceedings of the 14th International Conference on Research and Development in Information Retrieval*; October 1991, 32-45; Chicago, Illinois.

Crouch C.J. [1988]. "A Cluster-Based Approach to Thesaurus Construction." In: *Proceedings of the ACM Conference on Research and Development in Information Retrieval*; June 1988, 309-320; Grenoble, France.

Deogun J.S. and Raghavan V.V. [1988]. "Integration of Information Retrieval and Database Management Systems." *Information Processing and Management*, 24(3), 303-313.

Dillon M. and Gray A.S. [1983]. "FASIT: A Fully Automatic Syntactically Based Indexing System." *Journal of the American Society for Information Science*, 34(2), 99-108.

Fagan J. [1987]. "Experiments in Automatic Phrase Indexing for Document Retrieval: A Comparison on Syntactic and Nonsyntactic Methods." Doctoral dissertation, Cornell University, Ithaca, N.Y.

Fagan J. [1989]. "The Effectiveness of a Nonsyntactic Approach to Automatic Phrase Indexing for Document Retrieval." *Journal of the American Society for Information Science* 40(2), 115-132.

Fox C. [1992]. "Lexical Analysis and Stoplists." In: Frakes W.B. and Baeza-Yates R., (Ed.), *Information Retrieval: Data Structures and Algorithms*, Englewood Cliffs, NJ: Prentice-Hall.

Fox E.A., Nunn G.L., and Lee W.C. [1988]. "Coefficients of Combining Concept Classes in a Collection." In: *Proceedings of the ACM Conference on Research and Development in Information Retrieval*; June 1988, 291-308; Grenoble, France.

Frakes W.B. [1984]. "Term Conflation for Information Retrieval." In: *Proceedings of the Third Joint BCS and ACM Symposium on Research and Development in Information Retrieval*; July 1984, 383-390; Cambridge, England.

Frakes W.B. [1992]. "Stemming Algorithms." In: Frakes W.B. and Baeza-Yates R., (Ed.), *Information Retrieval: Data Structures and Algorithms*, Englewood Cliffs, NJ: Prentice-Hall.

Francis W. and Kucera H. [1982]. *Frequency Analysis of English Usage*, New York, NY: Houghton Mifflin.

Harman D. [1986]. "An Experimental Study of Factors Important in Document Ranking." In: *Proceedings of the ACM Conference on Research and Development in Information Retrieval*; September 1986, 186-193; Pisa, Italy.

Harman D. [1991]. "How Effective is Suffixing?" *Journal of the American Society for Information Science*, 42(1), 7-15.

Harman D. [1992a]. "Ranking Algorithms." In: Frakes W.B. and Baeza-Yates R., (Ed.), *Information Retrieval: Data Structures and Algorithms*, Englewood Cliffs, NJ: Prentice-Hall.

Harman D. [1992b]. "Relevance Feedback and Other Query Modification Techniques." In: Frakes W.B. and Baeza-Yates R., (Ed.), *Information Retrieval: Data Structures and Algorithms*, Englewood Cliffs, NJ: Prentice-Hall.

Harman D. [1992c]. "Relevance Feedback Revisited." In: *Proceedings of the 15th International Conference on Research and Development in Information Retrieval*; June 1991, 1-10; Copenhagen, Denmark.

Harman D. and Candela G. [1990]. "Retrieving Records from a Gigabyte of Text on a Minicomputer Using Statistical Ranking." *Journal of the American Society for Information Science*, 41(8), 581-589.

Katzer J., McGill M.J., Tessier J.A., Frakes W., and DasGupta P. [1982]. "A Study of the Overlap among Document Representations." *Information Technology: Research and Development*, 1(2), 261-274.

Lewis D.D. [1992]. "Feature Selection and Feature Extraction for Text Categorization." In: *Proceedings of the Speech and Natural Language Workshop*; Feb. 1992, 212-217, Harriman, N.Y. Morgan Kaufmann.

Lewis D.D. and Croft W.B. [1990]. "Term Clustering of Syntactic Phrases." In: *Proceedings of the 13th International Conference on Research and Development in Information Retrieval*; September 1990; 385-504; Brussels, Belgium.

Lennon M., Peirce D., Tarry B., Willett P. [1981]. "An Evaluation of Some Conflation Algorithms for Information Retrieval," *Journal of Information Science*, 3, 177-188.

Lovins J.B. [1968]. "Development of a Stemming Algorithm." *Mechanical Translation and Computational Linguistics*, 11, 22-31.

Luhn H.P. [1957]. "A Statistical Approach to Mechanized Encoding and Searching of Literary Information." *IBM Journal of Research and Development*, 1(4), 309-317.

Pacak M.G. and Pratt A.W. [1978]. "Identification and Transformation of Terminal Morphemes in Medical English, Part II." *Methods of Information in Medicine*, 17, 95-100.

Porter M.F. [1980]. "An Algorithm for Suffix Stripping." *Program*, 14(3), 130-137.

Salton G., and Buckley C. [1988]. "Term-Weighting Approaches in Automatic Text Retrieval." *Information Processing and Management*, 24(5), 513-523.

Salton G., and Buckley C. [1989]. "A Comparison between Statistically and Syntactically Generated Term Phrases." *Technical Report TR 89-1027*, Cornell University: Computing Science Department.

Salton G., and Buckley C. [1990a]. "Improving Retrieval Performance by Relevance Feedback." *Journal of the American Society for Information Science*, 41(4), 288-297.

Salton G. and Buckley C. [1991]. "Automatic Text Structuring and Retrieval: Experiments in Automatic Encyclopedia Searching." In: *Proceedings of the 14th International Conference on Research and Development in Information Retrieval*; October 1991, 21-31; Chicago, Illinois.

Salton G., Allan J. and Buckley C. [1993]. "Approaches to Passage Retrieval in Full Text Information Systems." In: *Proceedings of the 14th International Conference on Research and Development in Information Retrieval*; June 1993, 49-58; Pittsburgh, Pa.

Salton G., Buckley C. and Smith M. [1990a]. "On the Application of Syntactic Methodologies in Automatic Text Analysis," *Information Processing and Management*, 26(1), 73-92.

Salton G., Zhao Z. and Buckley C. [1990b]. "A Simple Syntactic Approach for the Generation of

Indexing Phrases." *Technical Report TR 90-1137*, Cornell University: Computing Science Department.

Salton G. and McGill M. [1983]. *Introduction to Modern Information Retrieval*. New York, NY: McGraw-Hill.

Sparck Jones K. [1972]. "A Statistical Interpretation of Term Specificity and Its Application in Retrieval." *Journal of Documentation*, 28(1), 11-20.

Sparck Jones K. and Jackson D.M. [1970]. "The Use of Automatically-Obtained Keyword Classifications for Information Retrieval." *Information Storage and Retrieval*, 5, 175-201.

Strzalkowski T. [1992]. "Information Retrieval Using Robust Natural Language." In: *Proceedings of the Speech and Natural Language Workshop;* Feb. 1992, 206-211. Harriman, N.Y. Morgan Kaufmann.

Turtle H. and Croft W.B. [1991]. "Evaluation of an Inference Network-Based Retrieval Model." *ACM Transactions on Information Systems*, 9(3), 187-222.

van Rijsbergen, C.J. [1975]. *Information Retrieval*. London: Butterworths.

Walker, S. and Jones, R. M. [1987]. "Improving Subject Retrieval in Online Catalogues." British Library Research Paper 24.

Weiss, S.F. [1970]. "A Template Approach to Natural Language Analysis for Information Retrieval." Doctoral dissertation, Cornell University, Ithaca, N.Y.

Chapter 14

THE ROLE OF LINGUISTIC
ANALYSIS IN FULL-TEXT RETRIEVAL

Amy J. Warner

INTRODUCTION

Although a few full-text databases have been around for a long time (both
Westlaw and Lexis have existed since the mid 1970's), the proliferation of these
sources has occurred since the mid 1980s. At that time, computer storage capacity
became large and inexpensive enough to consider offering a significant amount of
literature online in full-text form. Now, full-text databases are one of the fastest growing segments of the database market, with the number of offerings growing from a
few dozen in 1980 to thousands by the end of the decade [Tenopir and Ro, 1990].

The scenario now widely discussed in the literature includes a major role for full-
text databases. Surely their advantage—being able to access full documents directly
rather than just their surrogates—is indisputable. At the same time that the storage of
more and more of these files is an advantage, however, some disadvantages to searching full text also exist. The intellectual aspects of text retrieval have been reduced to
keyword and, sometimes, manually assigned key phrase searching. Furthermore, in
most operational systems, protocols proven useful in bibliographic databases, such
as proximity and Boolean operators, are less effective in full-text files.

The unique problems of full text relate to the sheer quantity of text available for
processing, as well as the important fact that full-text files often are not summarized
into abstracts and assigned index terms. In full-text databases, one cannot therefore
safely assume, as has been done fairly successfully with bibliographic databases, that
the appearance of a word or phrase signifies a relevant document.

How to solve the problems of full-text retrieval is hotly debated in the literature
on both experimental and applied information retrieval. A major theme in this debate
is how to obtain effective retrieval results from large quantities of full text that have
not been manually indexed and/or abstracted. Automatic and semiautomatic approaches can make use of statistical, linguistic and artificial intelligence (AI) techniques.

This chapter discusses the role of linguistic analysis in text retrieval, emphasizing the problems inherent in full-text databases. Although linguistics is deemed to be a fruitful area to study, a major assumption of this chapter is that successful information retrieval will depend on research that appropriately blends principles and techniques from several other fields, such as statistics, psychology, and artificial intelligence (AI), as well as linguistics, into an effective system design. Designing flexible and effective retrieval systems will depend on knowing as much as possible about each of them, as well as how they interact in retrieval system design.

RELEVANCE OF LINGUISTICS TO INFORMATION RETRIEVAL

The relationship between linguistics and information retrieval has been controversial. On the one hand, a general assumption has it that the preoccupation with textual material within information retrieval assures the position of linguistics as a key component in research and development. On the other hand, the amount of research using linguistics in information retrieval systems has always been small, and current commercial systems still use few automatic linguistic processing techniques.

The debate about the relevance of linguistics in information retrieval began more than 20 years ago. Those who advocated the connection pointed out the natural relationship between the two fields, as well as the belief that information retrieval can only progress if the role of language is more fully understood [Sparck Jones and Kay, 1973; Montgomery, 1972]. Those not in favor of linguistic approaches in IR argued that results do not justify the added complexity [Lancaster, 1972; Van Rijsbergen, 1979].

Those opinions, however, were formed at a time when few textual files were available to search and few systems existed through which one could access them. Furthermore, most textual databases were bibliographic files of document surrogates (authors, titles and abstracts), and they contained fewer records than they do today. Finally, these databases were primarily searched by skilled intermediaries, who knew how to use effectively the crude linguistic searching devices, such as truncation and proximity operators. But today's databases have increased in size; full text is now commonplace; and the inexperienced searcher, or end user, uses a variety of databases and information systems.

There are basically two ways that full-text databases can be searched in current commercially available systems. The first way uses a variety of structural operators, such as truncation symbols and proximity and Boolean operators, which have been available since the early days of online systems. The second way uses statistical ranking to order large sets of retrieved items by predicted relevance to the request, from most to least relevant [Tenopir and Ro, 1990].

All of these systems are, and will continue to be, quite useful. However, they are linguistically crude and burdened with two general problems:

1. They use a minimum of the *grammatical structure* found in the original document.

266

2. They continue to rely on surface matching of key terms; thus, there is little real exploitation of the *meaningful elements* of documents and queries.

Indeed, two of the major findings of full-text retrieval research are linguistic in nature. One is that structural devices, particularly Boolean operators, often retrieve an unacceptably large number of irrelevant documents [Tenopir, 1984]. Also, recall is rather low, due in part to the fact that it is "difficult for users to predict the exact words, word combinations, and phrases that are used by *all* (or most) relevant documents and *only* (or primarily) by those documents [Blair, 1990; p. 101]. Thus, it seems that room exists for further linguistic sophistication in current commercially available information retrieval systems.

Moreover, IR researchers and developers have long been advocating the development of a more interactive information retrieval system—one that would act much like a human colleague. This type of information system presents a clear and major role for linguistics. As Montgomery [1981] states:

> To simulate in some sense the assistance provided by a human colleague, an active information system must have at least three types of knowledge.
>
> 1. In order to communicate and receive information, the system must have *linguistic knowledge*—that is, knowledge of lexical items (words, phrases), grammatical categories (noun, verb), and grammatical relations (subject of, object of) representing the linguistic knowledge of relevant domains.
>
> 2. The system must have *extralinguistic knowledge*—knowledge of the entities, attributes, events, processes and relations comprising the information models for the relevant domain.
>
> 3. The system must have the *ability to use* such knowledge—that is, the knowledge of procedures for utilizing linguistic and extralinguistic knowledge to achieve a particular goal [p. 375].

APPROACHES TO LANGUAGE ANALYSIS

Various fields approach language analysis in one way or another. The two that seem to have the most potential for information retrieval are linguistic theory and natural language processing (NLP). They approach language analysis in fundamentally different ways, which can be grouped under four broad headings:

PURPOSE—the analysis goal, particularly whether it is intended to be for some useful purpose or whether it is an end in itself;

DATA—what counts as admissible data and how can the data domain be circumscribed;

ANALYSIS—what aspects of the data (syntax, etc.) need to be analyzed and how is the analysis to proceed (principled versus ad hoc);

EVALUATION—how to determine success or failure of the analysis.

Linguistic theory studies language and languages with the sole purpose of constructing a theory of their structure and functions at a given point in time and without regard to any practical applications that the investigation of language and languages might have [Lyons, 1981].

The PURPOSE of linguistic theory is twofold: to formulate adequate theories of the structure of individual languages (grammars) and to devise a general linguistic theory or linguistic metatheory which will characterize the structural properties shared by all human languages. Linguistic theory is rooted both in mathematics and psychology. It seeks to construct a set of rules for specific languages and structural properties common to all languages (a highly mathematical formalism) and to explain the innate human linguistic capacity (competence) by modeling what the native speaker knows about his language (psychology) [Langacker, 1972].

The DATA of linguistic theory consists of spoken utterances or written texts. Since language is open-ended both structurally and lexically, the data is, by definition, incomplete. Nevertheless, whereas the emphasis could be on building grammars that are as comprehensive as possible, these are the exception rather than the rule. Linguistic analyses are performed on relatively small sets of data that exhibit a small range of linguistic phenomena.

ANALYSES are made within particular linguistic theories and formalisms with the aim of achieving the following: describing the data in hand while making correct predictions about new data (thus independently motivating the theory); and making significant linguistic generalizations both about the data in hand and about new data [Langacker, 1972].

Different analyses within the same theoretical framework and competing theoretical frameworks are EVALUATED according to two notions: 1) internal adequacy, which refers to the compatibility of a theory with the data it purports to describe; and 2) external adequacy, which refers to the compatibility with universal tendencies, i.e., its "naturalness" [Langacker, 1972].

Natural language processing is that area of research and application that explores how natural language entered into a computer system can be manipulated and stored in a form that preserves certain aspects of the original [Harris, 1985]. NLP offers two approaches to the problem of linguistic analysis—either a weak or strong equivalence is assumed between human language use (performance) and computer processing of linguistic data. The weak equivalence approach seeks to develop particular algorithms for processing language and is characterized by a strong engineering approach. The strong equivalence approach seeks to characterize the underlying nature of a particular computation and its basis in the physical world; this viewpoint, often known as the cognitive approach, says nothing less than that a computer program is a model of human cognition [Winston, 1984]. Thus, the PURPOSE of natural language processing may be to produce a useful product or to build theories of cognition within a computational framework.

Both NLP approaches are interested in the details of the way actual linguistic DATA are processed in cases where they might be *used* by a person in actual linguistic performance. Thus, the range of data which must be accounted for is very large, as in linguistic theory, although a different way must be found to constrain the data domain in order to make program construction possible. Generally, two strategies are employed alone or in combination: the subject domain is limited (i.e., sublanguage analysis) or only certain components (i.e., syntactic, semantic, pragmatic) are processed, making some systems more "robust" than others.

ANALYSES are intended to be (at least) computationally tractable. In addition, theoretical motivation may be sought—major concerns include how much of what types of "knowledge" should be included and how that knowledge should be represented, partitioned, and stored in the system.

Thus, EVALUATION within the applied domain depends on whether the system works and whether it is efficient. In addition, the theoretical domain adds the criterion of psychological reality of the data structures and algorithms used.

Linguistic theory and natural language processing are similar and different in several fundamental ways. Linguistic theory emphasizes the capture of generalizations in a data set, which in fact is often constructed for the purpose of examining only a very limited set of phenomena. Natural language processing aims to process the language of a living corpus, usually constructed for some other purpose entirely.

The role of computer technology in the two fields also varies significantly. Researchers in natural language processing must make their processing algorithms computationally tractable, since their goal is either to equate successful computation with a human cognitive process or to engineer a practical working system. Linguistic theoreticians, on the other hand, may use computer processing, but only as a tool to test the accuracy of their grammar rules, which are constructed a priori within a particular theoretical framework.

Language analysis in information retrieval shows the effects of these two fields in two general approaches to IR system study. One approach, a natural language processing orientation, focuses on system design and performance. Here, one or more linguistic devices are implemented in a working system, and the effects of that device—measured by recall and precision—determined. Thus, for example, one might implement a grammatical phrase indexing algorithm, and test its effect on system performance. The second approach, reflecting the influence of linguistic theory, studies linguistic patterns in the data (i.e., documents and queries) of the retrieval system. The goal here is to capture structural or semantic information as a set of rules that can be used in later system development. An example of this type of approach might be a linguistic analysis and description of synonymous phrases in a given subject domain, with a view to producing a rule set which could generate them.

The first approach emphasizes system performance, and all retrieval system devices are evaluated according to the quality of the system output they produce. The second approach, on the other hand, makes no direct claim that the isolation and description of linguistic phenomena will improve system performance—it is generally assumed that this type of analysis is necessary, but there is no overt testing of this assumption. They are, in fact, attempting to produce the *linguistic knowledge* referred to earlier by Montgomery [1981].

LINGUISTIC PROCESSING IN FULL-TEXT INFORMATION RE-TRIEVAL SYSTEMS

Table 1 (adapted from Doszkocs [1986]) shows four linguistic levels and their counterparts in commercial IR, statistical IR, and IR which includes linguistic analysis [Lancaster and Warner, in press]. Processing at one or more of these four levels goes on in all kinds of textual databases, including full-text files.

Table 1. Comparison of levels of processing in commercial, statistical and linguistic methods

LINGUISTIC LEVEL	COMMERCIAL IR	STATISTICAL IR	LINGUISTIC IR
Morphological	Truncation symbol	Stemming	Morphological analysis
Syntactic	Proximity operators	Statistical phrases	Grammatical phrases
Semantic	Thesaurus	Clusters of co-occurring words	Network of words/phrases with semantic relationships
Pragmatic	Paragraph searching Search heuristics	Relevance feedback	Text processing into topics Expert search aids

In the *morphological* level, the structure of individual words is manipulated. For example, most commercial information retrieval systems use truncation to search for groups of words with the same stem. Thus, if one wants to search for information on "computers," the string COMPUT? will search for all variants of that word (e.g., compute, computer, computers, computational, and so forth). However, the user usually must explicitly supply truncation. In contrast, statistical IR systems automatically stem words, either by eliminating endings on words that correspond to entries in a suffix dictionary or by eliminating a fixed number of characters from the end of a word [Salton and McGill, 1983]. True linguistic processing performs an actual morphological analysis, which uses more general morphological rules to modify words in different categories. For example, the following word pairs conform to the same rule for verb/noun variation—use of -y for verbs and -ication for nouns:

amplify/amplification
codify/codification
qualify/qualification
simplify/simplification
verify/verification

A few experimental information retrieval systems incorporate a true morphological analysis [Niedermair, Thurmair and Buttel, 1984]. Most, however, including full-text systems, use some kind of stemming algorithm.

Some questions have been raised about the usefulness of stemming in IR, and these questions should continue with the automatic processing of increasing amounts of full texts. It has been assumed that stemming improves retrieval performance, based on the results of the Cranfield study, in which several linguistic devices were tested for their effectiveness and where only simple stemming showed much benefit [Cleverdon et al., 1966]. However, later studies have reopened the issue of whether stemming is a useful retrieval device [Harman, 1991]. Furthermore, some have questioned whether stemming should be *weak* or *strong*—that is, whether only inflections, such as singular/plural suffixes, should be removed, or whether stemmers also should remove derivational suffixes, which signal different parts of speech (e.g., compute [verb], computer [noun], computational [adjective]) [Walker and Jones, 1987]. These basic questions should be addressed in full-text retrieval systems as well.

Syntactic, or grammatical, information is employed whenever a phrase or other structural element is used to search free text. Searching devices of this type are considered particularly important in full-text databases, where simple keyword searching often retrieves too many irrelevant documents. This type of information is invoked frequently by users of commercial systems whenever they use proximity operators to search phrases (e.g., online(w)interfaces). However, once again, they must consciously supply these operators.

In statistical and linguistic IR, on the other hand, automatic processing replaces what the user once had to supply. In the statistical method, an example is the substitution of "statistical phrases" in queries in place of simple keywords. In linguistic analysis, grammatical phrases are extracted on the basis of an automatic syntactic analysis. Some syntactic analysis routines extract grammatical compound noun phrases as surface occurrences within text [Dillon and McDonald, 1983]. Others go further and actually do some grammatical derivation of expressions [Ruge, Schwartz and Warner, 1991]. A simple example illustrating these various approaches follows:

    input string: "There are card, book and online catalogs."
    Statistical approach:
      CARD BOOK
      BOOK ONLINE
      ONLINE CATALOGS
    Linguistic approaches:
      extraction of compound noun phrase—ONLINE CATALOGS
      derivation and extraction of noun phrases— CARD CATALOGS
                                   BOOK CATALOGS
                                 ONLINE CATALOGS

Much of the effort in this area concentrates on the derivation of grammatical noun phrases from input constructions that are notoriously hard to process because they are often highly ambiguous, including coordinate constructions and prepositional phrases.

Semantic information is used whenever words or phrases which are similar in meaning in some way are incorporated. For example: [Lancaster, 1979]

1a.	blood level	2a.	serum level
1b.	level of the blood	2b.	level of the serum

Each pair consists of a compound noun phrase (1a. and 2a.) followed by a syntactic variant (1b. and 2b.). The important point for this example is that the members of each pair contain the same basic keywords ("blood" in the first pair, "serum" in the second), while the two separate pairs use a completely different word to mean essentially the same thing. In an IR system that automatically searches for syntactic variants, a simple structural rule (i.e., transpose elements of certain compound noun phrases and create a noun modified by a prepositional phrase) could be used to retrieve documents containing these two variants.

To bring in the different lexical items used to mean the same thing ("blood" and "serum"), however, requires some kind of online dictionary or thesaurus that contains lexical items and semantic relationships. In commercial systems, users generally must use the thesaurus online or in paper form and key in all related expressions. Clusters of statistically related terms have been derived for some automatic systems, although, in general, the terms are grouped because they tend to co-occur and bear little true semantic similarity to each other. In contrast, linguistic approaches to semantics actually use a machine-readable thesaurus, often stored as a semantic network, which is then used as a mechanism for enhancing searches with semantically related terms. An example illustrates the difference between statistical and linguistic thesauri:

Statistical clusters	Linguistic thesaurus		
asymptotic	sheep	TAX	animal
criterion	sheep	TFOOD	grass
cycle			
problem	ram	MALE	sheep
	ewe	FEMALE	sheep
[Sparck Jones, 1971]	lamb	CHILD	sheep
	[Fox, et al, 1988]		

A common theme in full-text retrieval which incorporates semantics is the ability to consider the *concepts* referred to in texts, rather than simply their surface key terms. This is often referred to as "concept-based information retrieval," which can be achieved only by building and maintaining an additional knowledge base containing information about the meanings of terms as well as, sometimes, the semantic relationships among terms. Such a resource can take the form of a lexicon, listing terms and their grammatical and semantic features [Sager, Friedman and Lyman, 1987]; a retrieval thesaurus [Bernstein and Williamson, 1984]; or an AI data structure, such as a semantic network or a frame-based representation [Rau, 1987; Tong, 1988].

The *pragmatic* level refers to regularities that govern the choice of language in communication [Crystal, 1987]. Thus, there are rules of coherent and cooperative discourse that govern both written texts and the structure of interactions between individuals. Within full-text information retrieval, one might consider within-paragraph searching to be a crude pragmatic device. Furthermore, when a user employs a search strategy development technique—for example citation pearl growing, where terms found in a relevant document are incorporated and used in a subsequent version of the search strategy—one might also say this is the retrieval system analogy of a structured, cooperative dialogue. The statistical version of this is Salton's well-known relevance feedback technique, in which terms are reweighted automatically based on users' relevance judgments [Salton, 1968].

A system that automatically incorporates pragmatic information from linguistic analysis is able to process full texts and extract the general topics covered [Hahn, 1990]. Furthermore, a pragmatically-based full-text retrieval system would also be a flexible, adaptive expert system, which would store and manipulate a knowledge base of search heuristics that the system could use to advise the user in ongoing strategy development. It is at this point, where the heuristics of a human expert are invoked, that one is really talking about an AI application—a true expert system [Gauch and Smith, 1989].

A generalization about these four levels is that they proceed from processing of the *surface structure* to processing of the *deep structure* of the text. Surface processing only manipulates the input string into other structural variants using the same words, while processing at the deep structure level allows for the substitution of entirely different strings based on an analysis of the text's underlying structure or meaning.

Closely related to this is the discussion in natural language processing of the *depth* versus *breadth* tradeoff, which means that currently one can carry out sophisticated semantic and pragmatic processing only in very restricted subject domains. As the need to process a greater variety of text increases, analyses become more shallow because semantic and pragmatic processing can only be carried out with the addition of a knowledge base of linguistic meanings and uses—and this is only feasible for a limited amount of text.

CONCLUSION

This chapter has proceeded from the controversial premise that linguistic analysis can play an important role in full-text information retrieval research and development. There are two basic ways this can occur. One way uses linguistic analysis as one of a number of tools in engineering an effective retrieval system. Here, both linguistic and statistical processing are on equal footing, and the respective amounts of each, as well as how they should interact, must be part of any system implementation and evaluation [Cooper, 1984].

The other way in which linguistic analysis is relevant concerns the description of the textual data contained in information retrieval systems. We know comparatively little about the regular linguistic patterns of documents and queries on just about all the levels described previously. Thus, it is not surprising that applications of linguistics to information retrieval systems have yielded little improvement in system performance. Therefore, a basic research agenda should include the study of large amounts of text in information retrieval systems with the aim of discovering frequent, regular linguistic patterns that can be incorporated into the *linguistic model*, advocated by Montgomery [1981].

How far information retrieval can go with linguistic techniques remains an open question. The goal of a fully automatic, fully flexible retrieval system may never be realized, with or without the addition of a linguistic component. Nonetheless, systems can surely be made *more* flexible, adaptable and responsive than they are. And, in the process, we can also learn something about the linguistic structures inherent in texts and queries.

NOTES

Bernstein, Lionel M. and Williamson, Robert E. "Testing of a Natural Language Retrieval System for a Full Text Knowledge Base." *Journal of the American Society for Information Science*; 1984 July; 35(4): 235-247.

Blair, David C. *Language and Representation in Information Retrieval.* Amsterdam: Elsevier; 1990.

Cleverdon, Cyril; Mills, Jack; Keen, Michael. *Factors Determining the Performance of Indexing Systems.* Cranfield, England: College of Aeronautics; 1966.

Cooper, W.S. "Bridging the Gap Between AI and IR." In: Van Rijsbergen, C.J. (ed.), *Research and Development in Information Retrieval.* Proceedings of the Joint BCS and ACM Symposium; 1984 July 2-6; Cambridge, England. Cambridge, England: Cambridge Univ. Press; 1984: 259-265.

Crystal, David. *The Cambridge Encyclopedia of Language.* New York: Cambridge University Press; 1987.

Dillon, Martin; McDonald, Laura. "Fully Automatic Book Indexing." *Journal of Documentation*; 1983 September; 39(1): 135-154.

Doszkocs, Tamas E. "Natural Language Processing in Information Retrieval." *Journal of the American Society for Information Science*; 1986 July; 2(4): 364-380.

Fox, E. A. et al. "Building a Large Thesaurus for Information Retrieval." In: *Proceedings of the 2nd Applied Natural Language Processing Conference*; 1988 February 9-12; Austin, TX. Morristown, NJ: Association for Computational Linguistics; 1988: 101-107.

Gauch, Susan and Smith, John B. "An Expert System for Searching Full Texts." *Information Processing and Management*; 1989; 25(3): 253-263.

Hahn, Udo. "Topic Parsing: Accounting for Text Macro Structures in Full-Text Analysis." *Information Processing and Management*; 1990; 26(1): 135-170.

Harman, Donna. "How Effective is Suffixing?" *Journal of the American Society for Information Science*; 1991 January; 42(1): 7-15.

Harris, M.D. *Natural Language Processing.* Reston, VA: Reston Publishing; 1985.

Lancaster, F. Wilfrid. *Information Retrieval Systems: Characteristics, Testing and Evaluation*, 2nd ed. New York: Wiley; 1979.

Lancaster, F.W. *Vocabulary Control for Information Retrieval*. Washington, D.C.: Information Resources Press; 1972.

Lancaster, F. Wilfrid and Warner, Amy. *Information Retrieval Today*. Washington, D.C.: Information Resources Press; 1993.

Langacker, R.W. *Fundamentals of Linguistic Analysis*. New York: Harcourt, Brace Jovanovich; 1972.

Lyons, John. *Language and Linguistics: An Introduction*. Cambridge, England: Cambridge University Press; 1981.

Montgomery, Christine A. "Linguistics and Information Science." *Journal of the American Society for Information Science*; 1972 May-June; 23(3): 195-219.

Montgomery, Christine A. "Where Do We Go From Here?" In: Oddy, R.N. et al (eds.), *Information Retrieval Research*. London: Butterworth; 1981: 370-385.

Niedermair, G.T.; Thurmair, G.; Buttel, I. "MARS: A Retrieval Tool on the Basis of Morphological Analysis." In: Van Rijsbergen, C.J., (ed.), *Research and Development in Information Retrieval*. Proceedings of the Joint BCS and ACM Symposium; 1984 July 2-6; Cambridge, England. Cambridge, England: Cambridge Univ. Press; 1984: 369-381.

Rau, Lisa F. "Knowledge Organization and Access in a Conceptual Information System." *Information Processing and Management*; 1987; 23(4): 269-283.

Ruge, Gerda; Schwartz, Christoph; Warner, Amy J. "Effectiveness and Efficiency in Natural Language Processing for Large Amounts of Text." *Journal of the American Society for Information Science*; 1991 July; 42(6): 450-456.

Sager, Naomi; Friedman, Carol; Lyman, Margaret. *Medical Language Processing: Computer Management of Narrative Data*. Reading, MA: Addison-Wesley; 1987.

Salton, Gerard. *Automatic Information Organization and Retrieval*. New York: McGraw Hill; 1968.

Salton, Gerard; McGill, Michael J. *Introduction to Modern Information Retrieval*. New York: McGraw Hill; 1983.

Sparck Jones, Karen. *Automatic Keyword Classification for Information Retrieval*. London: Butterworth; 1971.

Sparck Jones, Karen; Kay, Martin. *Linguistics and Information Science*. New York: Academic Press; 1973.

Tenopir, Carol. "Retrieval Performance in a Full Text Journal Article Database." PhD dissertation, University of Illinois at Urbana-Champaign; 1984.

Tenopir, Carol; Ro, Jung Soon. *Full Text Databases*. New York: Greenwood Press; 1990.

Tong, R.M.; Applebaum, L.A. "Conceptual Information Retrieval from Full Text." In: RIAO '88: *User-Oriented Content-Based Text and Image Handling*; 1988 March 21-24; Cambridge, Massachusetts. Paris: C.I.D., 1988; 899-909.

Van Rijsbergen, C.J. *Information Retrieval*, 2nd ed. London: Butterworths; 1979.

Walker, S.; Jones, R.M. *Improving Subject Retrieval in Online Catalogues: 1. Stemming, Automatic Spelling Correction and Cross-Reference Tables*. London: British Library; 1987.

Winston, P. *Artificial Intelligence*, 2nd ed. Reading, MA: Addison-Wesley; 1984.

Chapter 15

TEXT BASED APPLICATIONS
ON THE CONNECTION MACHINE

Brij Masand, Stephen J. Smith and David Waltz

INTRODUCTION

Databases containing text are now being collected and generated so rapidly that processing and access systems cannot keep pace. Human processing is generally part of database creation and maintenance and, as data volumes increase, frequently becomes the bottleneck and/or dominant cost. Databases are balkanized, scattered across many different platforms and stored in incompatible formats. Access methods lack standards, are often difficult to learn and to use, and offer only mediocre performance.

But help is on the way. The cost of computing continues to drop dramatically—at a rate of a factor of ten about every five years. And progress is being made in using this increasingly affordable computer power for solving the problems of database maintenance and use.

This chapter reports on progress in several text-related tasks. All involve the Connection Machine supercomputer, a massively parallel machine that uses many relatively small processors to provide large amounts of power and memory at costs per operation similar to those of PCs and workstations, and much cheaper than mainframes or traditional supercomputers. Such systems are inherently scalable, and can grow as databases grow while running the same applications code.

What opportunities do supercomputers offer for text-based systems? An obvious answer is that we can process and provide user access to much larger databases and we can quicken response times. If cost were not an obstacle we could build a text retrieval system that would provide interactive (under ten-second response) to a ten terabyte database (see the "Text Retrieval" section following). We can also allow more complex queries, without causing unacceptable slowing of our systems. Thus, one central goal of our efforts is to provide systems that can scale up to arbitrarily large sizes without requiring frequent reimplementation of applications, and at the same time can operate interactively with a user.

Less obviously, we can provide friendlier systems (e.g. the WAIS and CMDRS systems described in "Text Retrieval") by trading computing cycles and memory for ease of use and ease of learning. And we can afford to use simpler algorithms that may be somewhat wasteful of computing cycles and memory, but that more than make up for this because they are simpler and cheaper to program (for example the automatic keywording and automatic Census return classification systems described later in the "Classification" section). Such methods can lead to lower overall system deployment costs (i.e. when hardware, software, and human costs are all taken into account).

Of course we have not simply used brute force in our work—there are also many clever ideas represented in what follows. Still, the main message depends critically on continuing advances in hardware power, and this is important: human costs will eventually dominate overall system costs, because no matter how small a fraction of total cost they are today, hardware costs will continue to plummet, while human costs are likely to rise. We cannot avoid human costs: if we do not change to new systems, humans will continue to be needed to perform critical roles in overall system operation; and if we want to change to novel architectures or add system functionality, human software engineering will be required. This exemplifies a second major goal of our work: to reduce human effort and costs—for users, for system designers, and for persons in the processing loop (editors, keyworders, data entry workers, etc.). This chapter provides evidence that our major goals can be achieved.

TEXT RETRIEVAL

The rapid growth of on-line databases is a great challenge to information retrieval technology. The Dow Jones News Retrieval service, for example, now offers the full text (for the last six months) for about 300 sources, and contains on the order of tens of gigabytes of text. Even the largest databases such as the Library of Congress (estimated at 40 terabytes) are becoming accessible through optical disk technology.

In this section we describe CMDRS (the Connection Machine Document Retrieval System), a production version of the text retrieval system reported by Stanfill and Kahle;[1] and then WAIS, a wide area information server that links thousands of text sources across several countries with a standard interface and protocol.

Connection Machine Document Retrieval System—CMDRS

The architecture of the Connection Machine (CM) is particularly appropriate for very fast text retrieval applications, provided that the database can fit into the CM's primary memory. A document is stored per processor and a search is performed by broadcasting query terms simultaneously to all the processors which, in parallel, compare the query terms to their own document.

For instance, in the query "Iran contra arms deal" the four query words are broadcast sequentially and their scores accumulated in the processors that have a document that contains that query term. After the scores are accumulated for the entire query, a ranked list of documents is returned. At this point, the user may look at the results of this first phase and decide that certain documents are more relevant than others. These specially marked "relevant" documents are fed back into the search by formulating a query from the searchable terms from all of these documents (potentially constructing a query as large as hundreds of query terms) that specifically represent the concept under examination. This second list of ranked documents may be closer to what the user is looking for.

The approach, called "relevance feedback," has been well known since the sixties but has only found practical use since the availability of, relatively, inexpensive supercomputers in the form of massively parallel machines. Using the Connection Machine CM-2, and compressing the text (see below) databases as large as 25 gigabytes can be easily searched.

Preprocessing the Text for Retrieval

The first step in preprocessing the text is compression. This is done by eliminating stop words (368 noncontent-bearing words such as "the", "on", "and") and then eliminating the most common words that account for 20 percent of the occurrences in the database. The second step removes a total of 72 additional words. The remaining words, known as *searchable terms*, are assigned weights inversely proportional to their frequencies in the database. Documents are represented using the surrogate coding method described by Stanfill and Kahle.[2] In this method groups of 25 words are collected and used to set bits in a 1024 bit vector using a hashing scheme. In searching a document the same hash codes that were used to set the 1024-bit vector are applied to each one of the query terms and the resulting hash bits are broadcast to all the processors to see whether a match exists. Although general phrases are ignored, pairs of capitalized words that occur more than once are recognized and are also searchable. More than 250,000 searchable words and word pairs occur in this database. Relevance feedback is performed by constructing queries from all the text of the document.

Disk-based Relevance Feedback Retrieval Systems for the Connection Machine

Stanfill and colleagues describe an algorithm that can potentially access a terabyte of information in less than ten seconds.[3, 4] This algorithm is applicable to databases ranging from 1 to 1000 gigabytes and uses an inverted index stored on disk, allowing a more flexible retrieval than was previously possible with the CMDRS signature-based approach. CMDRS is appropriate for large centralized databases. But text is available widely across organizations. What can be done with such distributed data?

Wide Area Information Server—WAIS

WAIS is an information retrieval system that capitalizes on the availability of both networked workstations and hundreds of text databases by providing a standard protocol for document query/retrieval over a network. Traditionally, accessing different databases requires learning a specific query protocol for each system. WAIS provides a solution to this, allowing the decentralized creation and maintenance of thousands of text sources, all accessible with the same interface/protocol.

WAIS Overview

The Wide Area Information Server project was started by Thinking Machines to construct an electronic publishing system for nonprofessional searchers. It is a client-server system using a standard protocol (based on NISO Z39.50).[5] To make the system friendly to end users, natural language is used as the query format, and the retrieved information can be anything from text to video or formatted records. This system has been in use for over five years in a variety of research and commercial environments.

WAIS consists of three components: a client program that provides the user interface; servers to do the indexing and retrieval of documents; and a standard protocol to transmit the queries and responses. The client and server are isolated from each other through the protocol so that any client can be used that is capable of translating a user's request into the standard protocol. Likewise, any server capable of answering a request encoded in the protocol can be used. In order to promote the development of both clients and servers, the protocol specification is public.

On the client side, questions are formulated in natural language. The client application translates them into the WAIS protocol and then transmits it over a network to a server. The server receives the transmission, translates the received packet into its own query language, and searches for documents satisfying the query. The list of relevant documents is then encoded into the protocol and transmitted back to the client. The client decodes the response and displays the results. The documents can then be retrieved from the server.

From the users' point of view, a server appears to be locally accessible but, in fact, could be in very different locations: on the local machine, on a network, or on a system connected by modem. The user's workstation keeps track of a variety of information about each server. The public information about a server includes how to contact it, a description of the contents, and the cost. In addition, individual users maintain certain private information about the servers they use. Users may need to budget the money they are willing to spend on information from particular servers, they need to know how often and when each server is contacted, and they need to assess the relative usefulness of each server. This information helps guide the workstation in making cost effective decisions in contacting servers.

With most current retrieval systems, complications develop as soon as one begins dealing with more than one source of information. The most common problem is that of asking a particular question. For example, one contacts the first source, asks it for information on some topic, contacts the next source, asks it the same questions (most likely using a different query language, a different style of interface, a different system of billing), contacts the next source, and so on. One of the primary motivations behind the initial development of the WAIS system was to replace this method with a single interface.

With WAIS, the user selects a set of sources to query for information and then formulates a question. When the question is run, the system automatically asks all the servers for the required information with no further interaction necessary by the user.

The documents returned are sorted and consolidated in a single place, to be easily manipulated by the user. Figure 1 shows how a user can select a source for an initial query with seed words, examine the resulting documents from the query, and choose one of the result documents for relevance feedback. The user has transparent access to a multitude of local and remote databases.

One of the most far-reaching aspects of this project is the development of an open protocol. Three companies (Thinking Machines, Apple, and Dow Jones) have jointly specified a standard protocol for information retrieval. Ideally this protocol would be internationally standardized, yet flexible enough to adapt to new ideas and technologies; and functional over any electronic network, from the fastest optical connections to telephone lines.

The use of an open and versatile protocol fosters hardware independence. This not only provides for a much wider base of users, it allows the system to evolve seamlessly over time as hardware technology progresses. It provides incentive to produce the best components possible. For example, the protocol provides for the transmission of audio and video as well as text, even though most workstations cannot yet handle those media. They are also free, however, to ignore pictures and sound returned in response to query, and to display and retrieve only text. This inability, though, does not hinder higher-end platforms from exploiting their greater processing power and network bandwidth.

The WAIS protocol uses the existing Z39.50 standard[6] as a base and extends it to incorporate many of the needs of a full-text information retrieval system. To allow future flexibility, the standard does not restrict the query language or the data format of the information to be retrieved. Nonetheless, a query convention has been established for the existing servers and clients. The resulting WAIS Protocol is general enough to be implemented on a variety of communications systems. We are working with the standards committee to put the needed functionality into a future version of the standard.

The success of a WAIS-like system depends on a critical mass of users and information services. In order to encourage development and use, Thinking Machines is not only publishing a specification for the protocol, but is also making the source code for a WAIS protocol implementation freely available. While this software is available at no cost, it comes with no support. We hope that it will help others in developing servers and clients.

CLASSIFICATION

Introduction

We now elaborate on the theme introduced before: how to use raw hardware power to reduce the human costs involved in producing software/systems for information retrieval. Traditionally, classifying or assigning keywords to documents has been performed by human editors. Automated systems have been developed, but for the most part these have relied on painstakingly constructed definitions of various

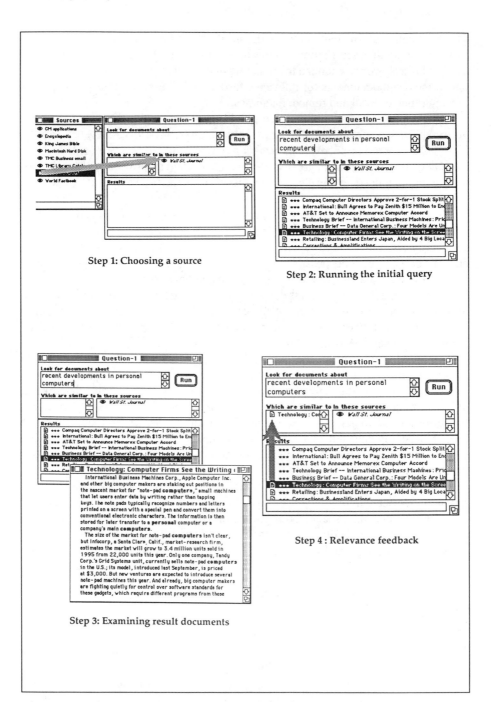

Step 1: Choosing a source

Step 2: Running the initial query

Step 3: Examining result documents

Step 4 : Relevance feedback

Figure 1. Document Selection Using the WAIStation

concepts from specific domains, requiring as much as several person-years of effort. In this section, we first take a look at the role of keywords/classification for accessing information and then describe two applications—the classification of Census returns and news stories—that demonstrate dramatic reductions in software engineering costs for automated classification, when given a preclassified database for training.

Indexing and Classification

Indexing can refer to at least two distinct activities. First it can refer to automated indexing of text documents where the documents are indexed for retrieval with respect to important words or phrases (terms) occurring in the text. In this sense of indexing, the indexing features (terms) need not represent a subject or a concept but rather allow the expression of a query (which may represent a concept) through such index features. Such indexing terms could be chosen on a statistical basis. The output of such an (automated) process might be a list of terms (along with their weights) and the documents that contain them [see Chapter 13]. Another interpretation of indexing may refer to arrangement of text documents into categories that do represent concepts or subjects in the domain of interest. In this case keywords or numbers representing concepts/subjects are assigned to each text document. In contrast to the work described in preceding sections where index features from text documents are used to facilitate full text retrieval and relevance feedback, this section describes the automated assigning of keywords that represent conceptual categories in the two domains of Census Classification and News Story Classification.

Example Full-Text Classification Problems

In many applications it is not enough simply to retrieve text: we would also like to be able to make decisions based on the content of the text. This section introduces two domains where such systems have been built, one for classifying news stories and another for classifying United States Bureau of Census long forms. Both tasks previously required expensive and time consuming human dependent classification systems and, as such, were ripe for automation.

The system built for the Bureau of Census classifies the long form responses of United States residents into approximately 800 different industry and occupation categories. The system built for Dow Jones Corporation is used to assign descriptive category codes to news articles from the Dow Jones Press Release News Wire. Both systems were built on the Connection Machine CM-2 and take advantage of large training databases of preclassified items. For this reason the systems did not require domain-specific hand coding of category definitions and, because of this, each system was developed in under two months.

Are Classification Systems Still Necessary?

In today's world of powerful online full text search and retrieval systems, it may seem somewhat anachronistic to be using category indexing. Nonetheless, many text databases in industry and government still include keywords, and many companies provide human keywording services as a major portion of their product. Why, however, would someone still care about such static and brittle systems when a dynamic search with such powerful operators as relevance feedback would enable them to interactively access the database? There are several reasons.

1. A priori classification is the low technology status quo

Consider customers of the Dow Jones news service who have been using the system for several years. They have spent significant effort in learning what the relevant keywords are for each document and what they mean. The system is not perfect and not always consistent but it is easy to understand once the keywords are known, and one need not be an expert in order to use such a system. It would be difficult to ask such users to now learn an interactive system.

2. An experienced editor can act as both a filter and a highlighter

In much the same way as the editors at a newspaper choose only a small fraction of the world's news each day for front page display, categorization systems can distill and mark for the user the information that is considered generally relevant to a topic. A good editor can also mark important pieces of information at the fringes of a topic.

3. Categorization provides a summary or a bird's-eye view of the data

By tracking the numbers of stories that are marked with particular keywords, one can get a good idea about the activity in various topics at different times. This is an important characteristic of preclassified databases for the news story classification project at Dow Jones. Similarly, one of the purposes of the classifications of Census long forms is to provide a picture of ways that the United States' population has shifted in occupation and in industry over time. With these assigned categories a member of Congress might easily get answers to questions such as: How many people are currently working in the auto industry? or, Is the computer industry still growing? Despite all the power of today's text retrieval systems, it would still be difficult to answer such questions by forming Boolean queries or using relevance feedback.

How Do We Measure the Performance of a Classification System?

As in most applications size, speed, and cost are important factors. Systems must be able to accommodate large amounts of information and to process it in a reasonable amount of time with an acceptable cost. Quality, as manifested in the consistency and completeness of classifications, is another important factor. If systems are not complete in their classifications, humans must finish the task, adding greatly to costs. Moreover, systems must respond not just to large and growing databases, but databases that change over time. Consider that just 10 years ago the category for AIDS did not exist; yet now there would be not only a category for AIDS but subcategories including such topics as research, politics, and education. These five aspects of classification systems (consistency, size, speed, change, and cost) are the most important to today's systems.

Speed

By definition "news" is fresh and dynamic. It has a much shorter information half life than, say, an encyclopedia article. Thus, speed is critical for any news article classification system. If it takes over a month to classify a story, most of the interest in that story may already have passed and the classification will not be used. Even in the Census task, the long form responses (though not really "news") had to be processed at a rate of 10 per second so that the 22 million returns could be categorized within the three months allowed by the schedule.

Cost

With the continued decrease in computer expense and the increase in cost of human labor it is clear that future computer systems will be far cheaper than human-dependent systems. This is provably true today for the Census classification task: we have demonstrated that the expensive Connection Machine CM-2 supercomputer can perform with the same consistency as human editors and yet provide a total cost savings over a human-based system.

Consistency

Today's classifications are nearly all assigned by human editors who are generally well trained, well educated, and relatively expensive. As such it can be time consuming and costly to classify articles and, because of the number of different people assigning categories, it is difficult to provide consistency. To achieve consistency, there must be review procedures to catch errors. Often a hierarchy of editors is formed where only a select few editors "define" the actual categories. It is, however, often difficult to define precisely a topic of interest or a category; "official" definitions of the categories generally allow individual editors considerable room for interpretation. At most 90 percent of the Census returns are correctly classified by the human editors, and the Dow Jones editors achieve only 83 percent recall and 88 percent precision when compared to an "expert" editor. Though consistency rates may seem low, the systems based on them are still useful. Human systems set the targets for acceptable levels of inconsistency and error for an automated system.

Change

Many of the databases being categorized today are dynamic. In the case of the Dow Jones Press Release News Wire, new categories are being created as world political and economic events change. Company names are a key portion of the categorization and these often change as companies enter or leave the market, and are merged or acquired.

Over the 10-year time span between censuses, completely new industries (such as genetic engineering) can appear or entire occupations (the neighborhood milkman) can disappear. Recognizing these changes is a main reason for collecting this database. Any useful categorization system needs to fit these changes by proposing new categories rather than trying to shoehorn responses into previously defined categories.

Classifying Census Long Forms

The Problem

As part of its mission to profile the people and the economy of the United States, the Census Bureau collects industry and occupation data for individuals in the labor force. For the 1990 Decennial Census, each of an estimated 22 million natural language responses to questions on the Census long form had to be classified into one of 232 industry categories and 504 occupation categories. If done fully by hand the cost of this task would be on the order of 15 million dollars. The industry and occupation data that the Census Bureau collects consists of free text and multiple choice responses from over 22 million citizens. The actual questions and a typical response look like this:

```
For whom did this person work?            Essex Electric
What kind of business or industry was     Photography-
this?                                      Battery Division
Is this mainly:  0. Manufacturing         0(Manufacturing)
                 1. Wholesale
                 2. Retail
                 3. Other?
What kind of work is this person          Apprentice
doing?                                     Electrician
What are this person's most important     Wiring Machinery
activities or duties?
Was this person employed by:              0  (Private
              0. private company               Company)
              1. federal government
              2. state government
              3. local government,
              4. self employed
              5. working without pay?
What is this person's age?                25
```

This response should be classified into the industry category "Photographic Equipment and Supplies" (code 380) and the occupation category "Electrician Apprentices" (code 576).

Before 1990, industry and occupation coding was performed using expensive and time consuming clerical methods. The clerks used bulky procedure manuals and dictionaries of phrases and codes to classify cases. A section of the alphabetical index to the industries coding manual that a clerk might use to classify the previous example is given below.

Phrase	Code	Industry Category
Photographic Apparatus	380	Photographic Equipment and Supplies
Photo. Cameras and Supplies	651	Sporting goods
Photo. Control Sys. - electronic	341	Radio TV and Communication Equipment
Batteries, automotive, secondhand	682	Miscellaneous Retail Stores
Battery manufacturing	342	Electrical Machinery, equipment
Battery - retail	620	Auto and Home Supply Stores
Battery - wholesale	500	Motor Vehicles and Equipment

An Automated Solution

An automatic classification system, PACE (Parallel Automated Coding Expert), was built for the Census to perform this classification task. The system runs on a massively parallel supercomputer and employs an empirical learning model called Memory-based Reasoning (MBR).[7-9] Following the MBR model, the PACE system uses a training database of 132,000 previously classified returns to classify new census returns not contained in the database. This contrasts with AIOCS (Automated Industry and Occupation Coding System), an automated system developed by the Census Bureau for the 1990 Census, which is essentially an expert system driven by knowledge extracted from human experts and tested via the same preclassified database. Interestingly PACE provides a more accurate, more robust and simpler solution than AIOCS though both systems were successful in lifting some of the classification burden from the human editors.

Memory-based Reasoning (MBR) was introduced by Stanfill and Waltz in 1986[10] and recent interest has been growing.[11, 12] Ideas related to MBR, however, have a long history.[13] MBR comprises a series of variations on nearest neighbor classification schemes, with the addition of other statistical techniques.[14, 15] The simplest nearest neighbor technique consists of assigning a given example to the same category as the preclassified example most similar to it. For example, a very simple version of MBR uses a hamming distance metric where the nearest neighbor is the example with the highest number of matching fields.

Although this idea is conceptually simple, it seems difficult to implement efficiently. MBR potentially requires comparing a given example to every training example in the database; and computing the similarity between examples and cases generally requires more sophisticated measures than hamming distance, since the likelihood is small of exactly matching free text strings. The larger the database of preclassified examples the more likely that the nearest neighbor algorithm will arrive at the correct classification, since an exact or nearly exact match becomes more probable. Storing and matching against these preclassified examples becomes more and more expensive, however, as the database increases in size. In general, classification accuracy increases slowly (less than linearly) as the database size increases, while the computational requirements of the larger database grow linearly.

The processing time can be reduced by using better algorithms. In fact nearest neighbor algorithms are in some ways very close to information retrieval algorithms used in text search engines. The information retrieval problem is usually solved on serial machines by use of a hash table or a precompiled inverted index or concordance. Text algorithms are particularly applicable to the census classification task since much of the relevant information in each example is contained in the free text fields of the examples.[16] Another approach to this match problem is to enhance speed by using parallel computing hardware where the number of processors (P) equals the number of preclassified examples (N), giving a nearly constant time solution. PACE uses a variant of this method where each processor holds 16 examples.

Memory-based Reasoning augments simple nearest neighbor match by weighting relevant data in the training examples more heavily than other data. For instance, consider the case where the industry field of a training example includes the following phrase: "The computer industry." Of the three words "computer" is clearly the most important. It could even be argued that "the" and "industry" occur so often in the database and in so many different categories that they should have little or no bearing in forming the match, as illustrated in the following case:

```
                                          Number of
                                   Exact Word Matches

Test Example:          "TheComputer Industry"


Training Example 1:"RetailComputer Sales"        1
Training Example 2:"TheAutomobile Industry"      2
```

If the hamming distance metric were used, then the second training example would be chosen as the nearest neighbor, though it should be obvious that the first example is a better match. Several feature weighting and evidence accumulation methods can correct this error. These weighting algorithms for MBR systems make use of information metrics similar to Shannon's[17, 18] or statistically derived metrics. Similar ideas on modified metrics have appeared in the nearest neighbor classification literature,[19, 20] but have not often been used in practice.

MBR can create useful features from fields in the original database data structure by grouping. For instance the conjunction of the Age and Industry Type fields may produce a nonlinear feature that is more useful at classification than either field by itself. In PACE, features are no more complex than two-element conjunctions. In principle, feature construction could be extended to n-element conjunctions.

Another addition to the simplest nearest neighbor methods is the use of k nearest neighbors for some decisions, rather than only using a single nearest neighbor. Our MBR model includes methods for combining information from the various weighted fields within an example, as well as algorithms for combining the information from the k nearest neighbors.

Evaluation Criteria

Two measures are relevant in evaluating the performance of these classification systems; we call them *accuracy* and *coverage*. Coverage measures the percentage of all returns/forms that the system attempts and are not referred to human coders. Accuracy measures the percentage of attempted classifications that are correct. The ideal system would thus have high accuracy (few mistakes) and high coverage (few referrals). In general both cannot be simultaneously maximized, and compromises must be made. In the case of the census system we were constrained by the problem requirement to produce a system that was at least as accurate as human coders. Thus this compromise required a decrease in the coverage in order to increase the accuracy to levels of 86 percent for occupation coding and 90 percent for industry coding. This tradeoff was accomplished by setting "referral thresholds" for each class, such that if the confidence score of the class of the best match does not exceed the threshold, the form is referred to a human coder.[21]

Cross validation was used to test and compare the different classification metrics. Both 2-way cross validation (where the database was randomly split into a test set and a training set) and n-way cross validation (where only a single test example was drawn from the database at a time) were used. In both cases the test and training set were kept completely isolated and the relevant statistics for feature weights were calculated solely from the training set. For the results that we cite, random test sets of at least 5000 examples were used. This is large enough to ensure the statistical significance of the estimates of overall accuracy and coverage.

Results

Following the MBR methodology, PACE uses a 132,247-example database for training and a disjoint set of examples as a test database. PACE matches each new census return with the entire training database and assigns a code based on the codes of the (previously classified) nearest matches. The 132,247-example database was also required for the development of the expert system as it was needed as a testing set to evaluate accurately system performance. Below we compare the performance of the two systems and their development times. The reported development time reflects the amount of effort to build each system assuming no existing development tools other than standard programming languages and directly compares the effort needed to develop each system exclusive of the time required by both systems to develop the training database. The performance numbers indicate the percentage of the database that can be classified automatically, while controlling accuracy to meet or exceed that of human coders (about 10% error rate on Industry Codes and 14% error rate on Occupation Codes).

	% of Industry Codes Assigned	% of Occupation Codes Assigned	Person-months Development Time Required
Human Editors	94	95	
AIOCS (expert based)	57	37	192
PACE (memory based)	63	57	4

These numbers show that PACE exhibits a 54 percent improvement over the expert system in coverage of occupation codes and a 10 percent improvement for industry codes. If AIOCS had been used in the 1980 Census processing, the expert system would have resulted in a 47 percent reduction in clerical workload and if PACE had been used, it would have resulted in a 60 percent reduction in clerical workload. These improvements are substantial given the size of the coding task for the 1990 Census (approximately 22 million returns) and its cost (approximately 15 million dollars).

Discussion

PACE, a memory-based reasoning classification system, has proven to be successful in performing the complex census coding task that has previously required human coders. This is also true of AIOCS from the Census Bureau, though PACE can accurately process 13 percent more of the returns than can AIOCS. The more important advantage of PACE is, we believe, the speed with which it was developed, nearly 50 times faster than AIOCS.

We attribute the significant decrease in PACE's development time to MBR's capability to exploit automatically the expert knowledge in a large, previously classified database, thus avoiding exhaustive hand coding of many rules. PACE, in turn, is made possible through the hardware and programming models of massively parallel supercomputers. The availability of this supercomputing power also allowed us to try many MBR variants and to test each extensively. This would not have been feasible on a slower machine.

Classifying News Stories

The Problem

Editors at Dow Jones Corporation each day assign codes to hundreds of stories originating from diverse sources such as newspapers, magazines, news wires, and press releases. Each editor must master the 350 or so distinct codes, grouped into six categories: industry, market sector, product, subject, government agency, and region. (See Figure 2 for examples from each category.) Due to the high volume of stories, typically several thousand per day, manually coding all stories consistently and with high recall in a timely manner is impractical. In general, different editors may code documents with varying levels of consistency, accuracy, and completeness.

Code	Name	# of Documents
R/CA	California	9811
R/TX	Texas	2813
M/TEC	Technology	9364
M/FIN	Financial	7264
N/PDT	New Products/Services	4149
N/ERN	Earnings	9841
I/CPR	Computers	2880
I/BNK	All Banks	2869
P/CAR	Cars	380
P/PCR	Personal Computers	315
G/CNG	Congress	307
G/FDA	Food and Drug Admin.	214

Figure 2. Some sample codes

The coding task consists of assigning one or more codes to a text document, from a possible set of about 350 codes.

An Automated Solution

Following the general approach of MBR, a system was built to perform the news wire classification task. The system proceeds by first finding the near matches for each document to be classified. This is done by constructing a relevance feedback query out of the document text, including both words and capitalized pairs, and sending it to the CMDRS document retrieval system. This query returns a weighted list of near matches (see Figure 3). Codes are assigned to the unknown document by combining the codes assigned to the k nearest matches (up to 11 nearest neighbors were used for this system). Codes are assigned weights by summing similarity scores from the near matches. Finally the best codes are chosen based on a score threshold. Figure 3 shows the headlines and the normalized scores for an example and first few near matches from the relevance feedback search.

Although MBR is conceptually simple, its implementation requires identifying features and associated metrics that enable easy and quantitative comparisons between different examples. A news story has a consistent structure: headline, author, date, main text, etc. Potentially one can use words and phrases and their co-occurrence from all these fields to create features.[22] For the purpose of this project we used single words and capital word pairs as features, largely because CMDRS, the underlying document retrieval system used as a match engine, provides support for this functionality.

Score	Size	Headline
1000	2k	Daimler-Benz unit signs $11,000,000 agreement for Hitatchi Data
924	2k	MCI signs agreement for Hitachi Data Systems disk drives
654	2k	Delta Air Lines takes delivery of industry's first . . .
631	2k	Crowley Maritime Corp. installs HDS EX
607	2k	HDS announces 15 percent performance boost for EX Series processors
604	2k	L.M. Ericsson installs two Hitachi Data Systems 420 mainframes
571	2k	Gaz de France installs HDS EX 420 mainframe
568	5k	Hitachi Data Systems announces two new models of EX Series mainframes
568	2k	HDS announces ESA/390 schedule
543	2k	SPRINT installs HDS EX 420
543	4k	Hitachi DataSystems announces new model of EX Series mainframes
485	4k	HDS announces upgrades for installed 7490 subsystems

Figure 3. Sample News Story with Eleven Nearest Neighbors

Results

The table below groups performance by code category for a random test set of 1000 articles. The last column lists the different codes in each code category. Although the automated system achieves fair to high recall for all the code categories, consistent precision seems much harder. Given CMDRS, the text retrieval system as the underlying match engine, we achieved these results in about 2 person-months. By comparison, Hayes[23] and Creecy[24] report efforts of 2.5 and 8 person-years, respectively, for developing rule/pattern-based concept descriptions for classification tasks with comparable numbers of categories. Our current speed of coding stories is about one story every two seconds on a 4k-processor CM-2 system.

Evaluation Criteria

For the results reported here we used n-way cross validation, which involves excluding each test example one at a time from the database and performing the classification on it. We used a randomly chosen set of 1000 articles for the test set.

Table 1. Performance for a Random Test Set

Cate-gory	Name	Recall	Precision	# of Codes
I/	industry	91	85	112
M/	market sector	93	91	9
G/	government	85	87	28
R/	region	86	64	121
N/	subject	72	53	70
P/	product	69	89	21
	Total	81	70	361

The performance reported here is the average for the entire database. It should also be possible to define confidence levels for an entire document so that only documents with a high confidence would be classified automatically with a high recall and precision, referring the difficult ones to the editors for manual coding.

The results are based on the assumption that the currently assigned codes are perfect, i.e. that all existing codes are appropriate and no appropriate codes are missing from any documents. The extra codes assigned by the automatic system are judged as inappropriate. It is natural to ask how complete or consistent the original codes are.

In order to judge the relevance of the extra codes and also to assess the consistency of coding, we asked the editors at Dow Jones to evaluate the codes assigned by the automatic system and also to re-evaluate the original codes that were assigned to the documents. The results for the first phase of this evaluation are described next.

Editorial Evaluation

Two hundred articles were selected at random. The articles were coded by the automatic system and the codes assigned were mixed with the original document codes, then sorted alphabetically to randomize them. These randomized codes (along with the text and without any scores) were evaluated by a single editor as *relevant* (correct), *irrelevant* (incorrect) and *borderline* (could be tolerated). The editor could also add for evaluation extra codes not on the list. The comparison below summarizes the results. Due to the small size of the evaluation set (200 articles), the evaluation results are suggestive rather than definitive.

We compared the original document codes (assigned earlier by editors) to the most recent code evaluation. Treating the *relevant* category of the editorial evaluation as correct and counting *borderline* codes as incorrect we find the following recall and precision for the original assignment:

Recall: 83%, Precision 88%

Including borderline codes as correct:

Recall: 61%, Precision 94%

This suggests that the editors are quite consistent in their coding and that they rarely assign *borderline* codes, the "maybe" category, which therefore should be treated as incorrect.

Comparing the performance of the automatic system considering only the relevant codes as correct (excluding borderline codes as incorrect):

Recall: 80%, Precision 72%

which is about the same as compared to the original codes assigned to the documents. This is not a surprising result, given the high consistency of editorial coding. Including borderline codes as correct:

Recall: 79%, Precision 73%

The automatic coding system does relatively well with respect to the *borderline* codes. Because the editors seldom assign them, however, it would seem better to filter *borderline* codes by improved confidence measures.

Discussion

A relatively simple MBR approach that uses the full-text of stories for determining similarity enables news story classification with good recall and precision for business-oriented news. While the performance seems less dramatic than certain systems that use manually constructed definitions (such as reports of 90 percent recall and precision[25, 26]), we believe that an MBR approach based on full-text comparison offers significant advantages in terms of ease of development, deployment, and maintenance. For instance, entirely new codes can be added either by including stories with the new codes into the database or by adding the new codes to some earlier stories in the database.

We should be able to improve system performance by increasing the size of the training database, since MBR systems benefit from larger databases.[27] Our test database can hold more than 120,000 stories on the existing hardware (about 50,000 were used for the above results).

Although we used an existing full-text, relevance feedback system as a match engine, we believe it would be relatively easy to build a match engine for this specific purpose. This approach can also be used to provide classification at little extra cost where a news retrieval system with relevance feedback already exists.

CONSEQUENCES AND FUTURE DIRECTIONS

This chapter has shown several ways that large amounts of compute power can be used to improve the lots of both end users of information and information application programmers. Systems we have built or prototyped have demonstrated faster response times (CMDRS), significant cost savings for implementation (Census application, and automatic keyword assignment system), as well as novel functionality

(WAIS). The methods that we have been using transfer to a wide range of important applications, including OCR (optical character recognition), medical diagnosis, insurance code assignment, and credit assessment. We believe that as the costs of computer hardware continue to drop dramatically and as high-speed networks continue to proliferate, applications of the sort described here will become widespread.

But many problems lurk just below the surface, and these will become evident as methods such as ours are applied to ever-larger databases. In the text retrieval area, we need to improve performance in order to deal successfully with 100-gigabyte and larger databases. In their 1985 article, Blair and Maron showed that STAIRS (or any other Boolean search system) achieves a recall-precision product of less than 20 percent.[28] A 100-gigabyte database of 1000-byte documents, where one document out of 1000 is actually relevant, would contain 100,000 truly relevant documents! To retrieve 75 percent of them (i.e., 75% recall), a user would have to access and examine 100,000 x 0.75/0.2 = 375,000 documents! CMDRS has been shown to do substantially better than Boolean search—its recall-precision product is on the order of 55 percent—but even so, a user would have to examine 100,000 x 0.75/0.55 = 136,364 articles. This suggests that the entire information retrieval paradigm used since the inception of the field needs radical rethinking.

The current MUC and TIPSTER efforts at DARPA (Defense Advanced Research Projects Agency) are predicated on turning text databases into more structured forms that can support question-answering instead of document retrieval. In these initiatives, frame-like structures, indexed under the topics (or events described) in individual sentences, with "slots" and "fillers" for people, companies, dates, quantities, etc. replace the original documents as the primary searchable database; the original text is appended to the frames. The frame structures allow specific answers to specific questions. For example, a request such as "Who is the president of Apple Computer?" would retrieve the correct answer, and would not return articles about presidents of companies that use Apple products or that are in the apple industry.

To date, most systems in this initiative have been largely hand-built. Clearly, hand-building will be inadequate for building frame templates to cover all topics in a 100-gigabyte database. Although some level of hand coding probably cannot be avoided,[29] we will need to develop, wherever possible, automatic or semi-automatic methods, using machine-readable dictionaries and clever automatic analyses of patterns in the text. In addition to the kinds of work we have done in automatic keywording of new articles (discussed earlier in this chapter), considerable progress has been made in recent years in using statistical methods for text analysis.[30]

Overall, however, it is important not to forget that computing power and memory per dollar spent continue to increase, and that maximum limits for high-end machines likewise are increasing. Technology advances are offering unprecedented opportunities for speeding up existing computations as well as for practical applications of ideas that were previously only of theoretical interest. These are exciting times.

NOTES

1. Stanfill, C. and Kahle, B. "Parallel Free-Text Search on the Connection Machine System." *Communications of the ACM* 29(12), 1229-1239.

2. Ibid.

3. Stanfill, C., Thau R., and Waltz, D. L. "A Parallel Indexed Algorithm for Text Retrieval." *SIGIR Proceedings, 1989* (Cambridge, MA.), 88-97.

4. Stanfill, C., and Thau R., "Information Retrieval on the Connection Machine: 1 to 8192 Gigabytes." *Information Processing and Management* 27(4), 285-310.

5. Z39.50-1988: *Information Retrieval Service Definition and Protocol Specification for Library Applications.* National Information Standards Organization (Z39), P.O. Box 1056, Bethesda, MD 20817. Telephone (301) 654-2512.

6. Ibid.

7. Stanfill, C. and Waltz, D. L. "Toward Memory-Based Reasoning." *Communications of the ACM* 29(12), 1213-1228.

8. Stanfill, C. and Waltz, D. L. "The Memory-Based Reasoning Paradigm." *Proceedings of the Case-Based Reasoning Workshop*, Clearwater Beach, FL, (May 1988), 414-424.

9. Waltz, D. L. "Memory-based Reasoning." In M.A. Arbib and J.A. Robinson (eds), *Natural and Artificial Parallel Computation*, The MIT Press, Cambridge, MA, (1990), 251-276.

10. Stanfill and Waltz, "Toward Memory-Based Reasoning."

11. Aha, D., Kibler, D. and Albert, M. "Instance-Based Learning Algorithms." *Machine Learning*, 6, Kluwer Academic Publishers, Boston, (1991), 37-66.

12. Atkeson, Chris. "Roles of Knowledge in Motor Learning." MIT AI Lab Technical Report 942, September 1986.

13. Dasrathy B.V. (ed.). *Nearest Neighbor (NN) Norms: NN Pattern Classification Techniques*, IEEE Computer Society Press, 1990. [This work provides an excellent review of nearest neighbor techniques.]

14. Waltz, D. L. "Massively Parallel AI." *Proceedings of the National Conference on Artificial Intelligence* (AAAI '90), Boston, (August 1990).

15. Waltz, "Memory-Based Reasoning."

16. Lorigny, J. Quid. "A General Automatic Coding Method." *Survey Methodology*, 14(2), 289-298.

17. Shannon, C. and Weaver, W. *The Mathematical Theory of Communication.* University of Illinois Press, Urbana, Ill, 1949.

18. Quinlan, R. "Learning Efficient Classification Procedures and Their Application to Chess End Games." In R. S. Michalski, J. Carbonell, and T. Mitchell (eds.), *Machine Learning: An Artificial Intelligence Approach*, Tioga Publishing, Los Angeles, CA, (1983), 463-482.

19. Short, R.D. and Fukunaga, K. "The Optimal Distance Measure for Nearest Neighbor Classification." *IEEE Transactions on Information Theory*, IT-27(5), 622-627.

20. Fukunaga, K. and Flick, T.E. "An Optimal Global Nearest Neighbor Metric." *IEEE Transactions on Pattern Analysis and Machine Intelligence*, PAMI-6(3), 314-318.

21. Chen, B.C., Creecy, R.H., and Appel, M.V. "On Error Control in Automated Industry and Occupation Coding," Proceedings of the American Statistical Association, Survey Methods Section (1991) (Note: This may be published in the *Journal of Official Statistics* by publication date of this chapter.)

22. Creecy, R. H., Masand B., Smith S, Waltz D., "Trading MIPS and Memory for Knowledge Engineering: Classifying Census Returns on the Connection Machine." *Communications of the ACM* 35(8), 48-63

23. Hayes, P. J. and Weinstein, S.P., "CONSTRUE/TIS: A System for Content-Based Indexing of a Database of News Stories." *Innovative Applications of Artificial Intelligence 2.* The AAAI Press/The MIT Press, Cambridge, MA, 1991, 49-64.

24. Creecy, et al.

25. Hayes and Weinstein.

26. Rau, Lisa F. and Jacobs, Paul S. "Creating Segmented Databases From Free Text for Text Retrieval." *SIGIR Proceedings, 1991* (Chicago, IL), 337-346.

27. Creecy, et al.

28. Blair, D.C. and Maron, M.E. "An Evaluation of Retrieval Effectiveness for a Full-Text Document-Retrieval System." *Communications of the ACM* 28(3), 289-299.

29. Lenat, G., Prakash, M., & Sheperd, M. (1986) "CYC: Using Common Sense to Overcome Brittleness and the Knowledge Acquisition Bottleneck." *AI Magazine*, 4, 65-85.

30. Church, K. "A Stochastic Parts Program and Noun Phrase Parser for Unrestricted Text". Unpublished manuscript, AT&T Bell Labs, Murray Hill, NJ, 1986.

Index

Coach expert searcher, 170

cognitive process of indexing, 165, 230-231

compatibility, *See* protocol standards

comprehensive searching, *See* broadening search

computer documentation
> online help, 91-101
> user needs, 92-93

computer-assisted indexing, 159-160, 163 *See also* automated indexing
> Chemical Abstracts Services, 221-222, 237
> future uses, 212
> indexer reactions, 185
> interface, 166-169, 181, 185-186, 236
> logical steps (NASA system), 219
> MedIndEx, 166-169
> online resources, 165, 234-237
> Pedernales (PDL) system, 177, 180-186
> prompts, 167-169, 236
> specificity, 206
> system components, 202
> validating terms, 181

Connection Machine
> CM, 277, 278
> CM-2, 283

Connection Machine Document Retrieval System, 278-279, 291-292

connections (chain of links), 129-131

consistency, *See* indexing quality

controlled vocabulary, *See also* thesauri
> vs. free-text 241
> vs. full-text 242-243

conventional indexing, *See* human indexing

coordinate indexing, 163-164

cost
> computer and human, 285
> system development, 295

coverage (evaluation measure), 289

cross-validation, 289, 292

D

DARPA, 295

data structure, *See also* record unit
> data relationships in art history, 32-33
> in frames, 166
> in full-text, 35
> Information Structure Management System, 118-134
> Witt/Census systems, 42-55

data types
> images, 15-17, 69-70
> metadata, 131-133
> relationships, 120-123
> scripts, 118-119, 129

date indexing, 30

descriptors, *See also* controlled vocabulary *or* thesauri
> as links 151

dictionary, online
> for linguistic analysis, 272-273
> Petroleum Abstracts Dictionary, 177-179, 181-184

digital vs. analog images, 69-70

display *See also* interface
> based on user input, 127, 139
> based on user knowledge, 149-150
> of thesaurus, 181-186, 233

document analysts, *See* indexing *or* abstracting

document similarity, *See* similarity

Dow Jones news classification project, 290-294

E

evaluation
> of classification systems, 284-285, 289
> of Dow Jones news classification system, 292-293
> of indexing, 222-230
> of MedIndEx, 172

expert systems 111, *See also* knowledge-based systems

explicit vs. implicit links, 80, 104, 106

explosive searching, *See* broadening search

301

SPECIALIST, 172
lexicon for linguistic analysis, 272-273

L

language processing, *See also* linguistic analysis
  disambiguation, 207
  isolating semantic units, 202-203
  phrase delineation, 205-206, 258-259
  punctuation, 250-251
  single terms vs. phrases, 209-211
  stemming, 252-254, 270-271
  supercomputers, 277-278
  vs. linguistic analysis, 267, 269
  word boundaries, 249-251
lexicon, *See also* thesauri
  for linguistic analysis, 272-273
library use, *See* information-seeking behavior
limiting searches, 149
linguistic analysis, 265-276, *See also* language processing
  in information retrieval systems, 270
  morphology, 270-271
  pragmatics, 273
  role in information retrieval, 266
  semantics, 272-273
  syntax, 271-272
  use of lexicon, 272-273
  vs. natural language processing, 267, 269
links
  as index entries, 79, 103-109, 150-151
  as pointers, 154
  as search criteria, 127-128
  between frames (slots), 166
  computed, 82
  connections, 129-131
  definition, 121
  document analysis, 86
  explicit vs. implicit, 80, 104, 106
  functions, 154
  in hypertext, 79-80, 81-82, 106-107
  in Unified Information Structure Management System, 120-123

inheritance, 126-127
management, 81-83
temporal, 82
types, 81-82, 107, 121-123, 132-134, 135
weighting, 121
Lister Hill National Center for Biomedical Communication, 161
Lovins stemmer, 252-253

M

machine-aided indexing, *See* computer-aided indexing
MAI, *See* NASA: Machine-Aided Indexing system
maintenance
  of knowledge-base, 211-212
  of thesauri, 191-192
  retrospective indexing, 192
mapping
  between thesauri, 204
  to controlled vocabulary, 203
MARC Visual Materials Format, 15
matching, *See* similarity
MBR (Memory-based Reasoning), 287-290
Medical Subject Headings (MeSH) in MedIndEx, 161-172
MedIndEx, 161-172
MEDLINE, 161-170
Memory-based Reasoning, 287-290
MeSH (Medical Subject Headings) in MedIndEx, 161-172
metadata, 131-133
morphology (language processing), 270-271
MUC, 295
multimedia „*See also* visual images, *or* hypermedia
  indexing possibilities, 99-101

N

NASA (National Aeronautics and Space Administration)
  Machine-Aided Indexing, 201-219
  Thesaurus, 187, 192

R

rank ordering, 149, *See also* weighting
    problems, 254-255
record unit
    defining, 248-249
    visual images, 17
relationships, *See also* links
    thesaural, 178, 179
relevance feedback, 257-258
result neighborhoods, 150
retrieval, *See* searching
retrieval/query protocol, 279

S

"S" stemming algorithm, 252
scanning, 26, 95, 97
schema neighborhoods, 127
scripts, 118-119, 129
searching, *See also* queries
    aids, 149-150, 170-171, 243-244,
    257, 267
    based on text structure, 148-149
    Boolean, 142-147, 266-267, 295
    broadening search, 125-126, 257-258
    display specification, 139
    document links, 138-139
    for visual images, 17
    full-text, 242-244, 265, 266-267
    hypermedia vs. traditional system,
    112-118
    interface, 116-117, 170-171
    multiple data sources, 280-281
    online indexes, 96, 98
    process, 112-113 117-138
    rank ordering, 149
    search types, 139-150
    serendipitous finds, 26, 36, 95, 97
    standardized query protocol, 279-281
    starting objects, 115-116, 138, 139-
    142
    target objects, 116, 138, 144-147
    versioning, 122, 149-150
    weighting criteria, 138

semantic analysis, 202-203, 272-273
semantic units, *See* language processing
serendipitous finds, 26, 36, 95, 97
similarity
    of document and query, 142-143, 147-
    148, 255-257
    of documents, 287-288, 291
single terms vs. phrases, 209-211, 254
Single-Word Assessment Tool, 209, 211
situation-specific assistance, 167-169
slots (relation links), 166
SPECIALIST, 172
speed
    faster algorithms, 288
    information availability, 285
    parallel processing, 278, 288
standards, *See* protocol standards
stemming, 252-254, 270-271
    limitations, 207, 271
stopwords, 216-218, 251-252
structure, *See* data structure
suffixing, *See* stemming
supercomputers, 277-278
    Connection Machine (CM), 277, 278
    Connection Machine (CM-2), 283
syntactic parsing, 259
syntactic tagger and bracketer, 259
syntax (language processing), 271-272

T

term relationships in thesauri, 178, 179
text processing, *See* language processing
thesauri
    automated construction, 212, 258
    automated text analysis, 272-273
    for visual images, 17
    in computer-aided indexing, 203,
    204
    NASA Thesaurus, 187
    online, 181-186, 233
    online vs. print, 191
    Petroleum Abstracts Dictionary, 177-
    179, 181-184

term relationships, 178, 179
term selection, 191
TIPSTER, 295
training indexers and abstractors, 232-234
translation
  between thesauri, 204
  to controlled vocabulary, 203
truncation, *See* stemming
typed links, *See* links: types

U

University of Maryland Historic Textile Database, 68
updating, *See* maintenance
user aids, *See* searching: aids
user modifications
  in hypertext, 81
  of access points, 150
  of data, 70, 118
user needs, *See also* versioning
  humanists vs. scientists, 23
  online indexes, 92-93, 95, 97
  visual images, 7-13

V

validation of input, 181, 192
versioning, 122, 149-150
virtual documents, 129, 134
visual images, 3-5, 7-22, 57-72
  description challenge, 4-5, 13, 17, 63, 69
  existing databases, 7, 63, 68

future possibilities, 70
indexing, 13
objective vs. subjective elements, 13, 17, 63
record content, 15-17, 66-67
storage (analog vs. digital), 69-70
subjective nature, 63
vs. text, 12-13
user needs, 20-22, 69
vocabulary, 17, 64-69
Visual Materials Format (MARC record), 15
vocabulary, *See* thesauri

W

WAIS (Wide Area Information Server), 279-281
  interface, 280, 282
weighting, *See also* rank ordering
  by computer, 142-143, 254-257, 260, 279, 291
  by indexer, 121, 181, 191
  by searcher, 138
  for classification, 291
  of descriptor importance, 181, 191
  of link importance, 138
  of link strength, 121
  of terms in text, 260, 279
  usefulness, 254-255
weighting for searching, 279
Wide Area Information Server (WAIS), *See* WAIS
Witt Computer Index, 27-55
word delineation, 249-251, *See also* phrase delineation